DEVELOPMENT FROM WITHIN

The last decade has been one of crisis for Africa. There is a general recognition that neither African governments nor foreign aid agencies have made a significant impact on the quality of life for rural peoples. Given the enormous variety of Africa – social, economic, political, cultural, environmental – the search for universal, macro-level solutions is inappropriate. Specificity is critical. A richer and more complex picture emerges from micro-level studies of the strategies which African rural peoples are using to survive the current crisis. These are economic, social, ecological and political in nature and show a remarkable degree of dynamism and innovation in the face of increasing pressures.

Development from Within focuses on the relationship between local action and macro-event, arguing for flexibility as an ideology as well as a strategy. Presenting an alternative framework for analysis, indigenous African scholars use case studies to explore the complex social relationships of power. Against the current tide of pessimism, the authors argue for the knowledge and skill of African peoples and illustrate the diverse means by which men and women in rural Africa struggle to survive.

Both editors have worked in Africa and are members of the Department of Geography at Carleton University, Ottawa. **D.R. Fraser Taylor** is co-author of the influential *Development from Above or Below*. **Fiona Mackenzie** has recently completed an in-depth historical analysis of agricultural and rural change in a smallholding area of Kenya. The authors of the case studies represent a new generation of dynamic African scholarship.

42 - 361

DEVELOPMENT FROM WITHIN

SURVIVAL IN RURAL AFRICA

Edited by
D.R.F. Taylor and Fiona Mackenzie

London and New York

First published 1992
by Routledge
11 New Fetter Lane, London EC4P 4EE

Simultaneously published in the USA and Canada
by Routledge
a division of Routledge, Chapman and Hall Inc.
29 West 35th Street, New York, NY 10001

© 1992 D.R.F. Taylor and F. Mackenzie

Typeset in Baskerville by
Pat and Anne Murphy, Highcliffe-on-Sea, Dorset
Printed in Great Britain by
Biddles Ltd, Guildford and King's Lynn

British Library Cataloguing in Publication Data
Development from within: survival in rural Africa.
1. Africa. Rural Regions. Economic conditions
I. Taylor, D.R. Fraser (David Ruxton Fraser) 1937–
II. Mackenzie, Fiona
330.96
ISBN 0–415–03567–8
ISBN 0–415–03991–2 pbk

Library of Congress Cataloging in Publication Data

Development from within: survival in rural Africa / edited by
D.R.F. Taylor and Fiona Mackenzie.
p.   cm.
Includes bibliographical references and index.
ISBN 0–415–03567–8 – ISBN 0–415–06991–2
1. Rural development – Africa, Sub-Saharan – Case studies.
2. Community organization – Africa, Sub-Saharan –
Case studies.
3. Cooperative societies – Africa, Sub-Saharan – Case studies.
I. Taylor, D.R.F. (David Ruxton Fraser), 1937–
II. Mackenzie, Fiona.
HN780.Z9C643   1992   307.1′412′0967–dc20
91-10050 CIP

722758

CONTENTS

TABLES

FIGURES

CONTRIBUTORS

Noel A. Chavangi, Researcher, formerly with the Kenya Woodfuel Development Programme, Nairobi, Kenya.

George J.S. Dei, Assistant Professor, Department of Sociology and Anthropology, University of Windsor, Ontario, Canada.

Fiona Mackenzie, Associate Professor, Department of Geography, Carleton University, Ottawa, Ontario, Canada.

Takyiwaa Manuh, Research Fellow, Institute of African Studies, University of Ghana, Legon, Ghana.

Japheth M.M. Ndaro, Director, Institute of Rural Development Planning, Dodoma, Tanzania.

Alice Nkhoma-Wamunza, Acting Director, University Library Services, University of Dar es Salaam, Tanzania.

Peter O. Ondiege, Lecturer, Department of Urban and Regional Planning, University of Nairobi, Kenya.

Jacob Songsore, Senior Lecturer, Department of Geography and Resource Development, University of Ghana, Legon, Ghana.

D.R.F. Taylor, Associate Dean (International), Faculty of Graduate Studies and Research; Director, Carleton International; Professor, Department of Geography and School of International Affairs, Carleton University, Ottawa, Ontario, Canada.

Lovemore M. Zinyama, Senior Lecturer, Department of Geography, University of Zimbabwe, Harare, Zimbabwe.

PREFACE

In 1981 one of the editors (Taylor) was involved in the publication of *Development from above or below?: the dialectics of regional planning in developing countries* (Stöhr and Taylor 1981). The arguments for development 'from below' are now fairly well known. Succinctly, development 'from below' considers

> development to be based principally on maximum mobilization of each area's natural, human, and institutional resources with the primary objective being the satisfaction of the basic needs of the inhabitants of that area. In order to serve the bulk of the population broadly categorized as 'poor', or those regions described as disadvantaged, development policies must be oriented directly towards the problems of poverty, and must be motivated and initially controlled from the bottom. There is an inherent distrust of the 'trickle down' or 'spread effect' expectations of past development policies. Development from below strategies are basic-needs oriented, labour-intensive, small-scale, regional-resource-based, often rural-centred, and argue for the use of 'appropriate' rather than 'highest' technology.
>
> (Stöhr 1981: 1–2)

As originally expressed, 'development from below' had several weaknesses. Regional political economy critics drew attention to the neo-populist or Utopian origins of the paradigm and its lack of adequate theoretical underpinning. In essence, the problems emanated from the lack of theorizing of social relations with respect to the composition of the State, the existence of class and gender differentiation, and, in turn, State–local relations. Writers such as Charles Gore (1984), for example, showed how problematical is the assumption that regional planning ('from above' or 'from below') takes place in a political vacuum. Regional planning, he proposes, is one of the instruments at the disposal of those who wield power within the State and can be understood only as an integral component of State policy. It may be used to legitimate government authority; chameleon-like, it may also be used to promote the material interest of those in power (Gore 1984: 242–58).

A lack of theorizing of the State in turn fed into a conception of society as undifferentiated or homogeneous. It was assumed, for example, by Stöhr (1981) that, by definition, local communities had commonly held interests and would act together in these interests. But the conflation of 'place' with 'people' in terms of identifying interest groups is problematical for the refinement of theory which seeks to be relevant to the diverse realities within Africa. Finally, the degree of conflict or confrontation that is likely to emerge in the event of significant mobilization locally, whether with local authority or that of the State, was ignored.

In all three of the above areas, substantial rethinking is necessary to ensure the development of theory that has greater historical and social specificity within Africa, as Hyden (1983) calls for. This can emerge only through an iterative process whereby small-scale event (at the intrahousehold level and through local community action) is examined in the context of large-scale process. It was in order to explore this relationship, specifically to focus on local-scale organization ('from within') in the context of economic crisis in Africa, that this book has been produced.

Through an analysis of case studies at the local level (all too frequently ignored as theorists measure success through the applicability of universals), which are set within the political and economic context of the State, the underlying objective of the book is to focus on the relationship between local action and macro event. The intent is to move beyond the rhetoric that has come to surround the concept of 'development from below', which is adopted now in the discourse of institutions such as the World Bank, to examine issues raised by earlier work. Drawing from the case studies, the aim is to suggest avenues for future theoretical exploration. Fundamentally, the book illustrates the diverse means by which women and men in rural Africa struggle to survive.

In the first chapter, Mackenzie examines alternative analyses of and prescriptions for the deepening crisis in Africa. Her concern is to expose questions which arise from such analysis with respect to the potential for local organization to effect positive change in rural Africa.

This sets the context for the following eight chapters, presenting eight case studies which are written from very different theoretical perspectives. The first of these is Zinyama's study of 'Local farmer organizations and rural development in Zimbabwe'. This chapter is particularly interesting because it argues that there is a very positive relationship between the State and local organizations.

Next are three case studies from Ghana, each of which gives a very different perspective on local initiatives. In Chapter 3 Dei describes the response of a community in southern Ghana to the growing stresses facing the people in the 1980s. Here the community appears to have acted as an entity and there is no evidence in the study of differentiation. This is quite different from Chapter 4 by Songsore, who discusses the co-operative credit movement in northern Ghana and shows that differentiation and conflict clearly exist. This is true also in the third case study from Ghana by Manuh. In Chapter 5 she documents the continuing political struggles and tensions of the salt co-operatives in Ada District. These demonstrate the complexities of intra-community relationships and the ambivalent relations among the State, local government, traditional authorities and the local people.

Two case studies are drawn from Kenya where local organization is largely seen as an extension of State planning. In Chapter 6 Ondiege looks at the situation in Machakos District and in Chapter 7 Chavangi considers some of the experiences of the Kenya Woodfuel Development Programme, drawing much of her material from experiences in Kakamega District. In both of these case studies the relationships between government and the local people appear to be quite different from the situation in Zimbabwe and Ghana and non-government organizations (NGOs) and international aid agencies are major actors. The case studies conclude with two studies from Tanzania. In Chapter 8 Ndaro looks at the experience of Dodoma District and argues that local initiatives will remain marginal to the development process unless they are integrated with the planning efforts of government. In Chapter 9 Nkhoma-Wamunza provides a fascinating description of the struggle which faced the women of the village of Utengule Usangu in Mbeya District who were attempting to organize beer-brewing co-operatives. Here, gender oppression was a major factor in undermining their efforts.

The book concludes with a chapter by Taylor which attempts to synthesize theory and practice relating to 'development from within' in Africa.

In *Development from above or below?*, it was argued that:

> The validity of development approaches will not be determined as a result of theoretical and ideological debate, but in the realm of practice. The peasant families of Africa . . . are more likely to judge the validity of a strategy from its results rather than its ideological or methodological soundness.
>
> (Stöhr and Taylor 1981: 458)

It is clear from empirical evidence that existing strategies, whether socialist or capitalist, have not brought the results expected of them. Sandbrook has posed the question:

> Is it wholly unrealistic to hope that, out of systemic crisis and popular disillusionment and withdrawal can emerge the self-reliant and communitarian basis for the construction of more organic and satisfactory economic and political structures? Or are statist top-down forms of capitalism and socialism the only practicable frameworks for economic development today?
>
> (Sandbrook 1986: 331)

Development 'from within', like development 'from below',

> argues for flexibility and is as much an ideology as a strategy. It is a way of looking at development rather than a rigid set of policies and ideas. In practice there will be many responses to it over both time and space.
>
> (Stöhr and Taylor 1981: 458)

Robert Chambers (1989b: 24) has argued for an 'ideology of reversals' which starts with the priorities and conditions of rural people. Development from within is an example of such an approach. Both editors hope that this book will make a contribution to a greater understanding of the realities facing the people of rural Africa and that out of such an understanding will come both new theory and practice which are based on 'development from within'.

D.R.F. Taylor and
Fiona Mackenzie

ABBREVIATIONS

AAF–SAP	African Alternative Framework to Structural Adjustment Programmes
ACCOSCA	African Confederation of Co-operative Savings and Credit Association
AFC	Agricultural Finance Corporation
APPER	African Priority Programme for Economic Recovery
ASAL	Arid and Semi-Arid Lands
ATAF	Ada Traditional Area Fund
BSACo	British South Africa Company
CBS	Central Bureau of Statistics
CCA	Canadian Co-operative Association
CDR	Committee for the Defence of the Revolution
CDTF	Community Development Trust Fund
CEG	Commonwealth Expert Group
CIRDAFRICA	Centre on Integrated Rural Development for Africa
CMB	Cotton Marketing Board
CUA	Credit Union Association
DDC	District Development Committee
DFRD	District Focus for Rural Development
ECA	Economic Commission for Africa
ECOWAS	Economic Community of West African States
EEC	European Economic Community
EIU	Economist Intelligence Unit
ERP/SAP	Economic Recovery Programme/Structural Adjustment Programme
FAO	Food and Agriculture Organization
FCS	Farmers' Co-operative Societies

GATT	General Agreement on Trade and Tariffs
GDP	Gross Domestic Product
GMB	Grain Marketing Board
GNP	Gross National Product
IDS	Institute of Development Studies
ILO	International Labour Organization
IMF	International Monetary Fund
KVIP	Kumasi Ventilated Improved Pit
KWDP	Kenya Woodfuel Development Programme
LPA	Lagos Plan of Action
MIDP	Machakos Integrated Development Programme
NBC	National Bank of Commerce
NFAZ	National Farmers' Association of Zimbabwe
NGOs	Non Government Organizations
OAU	Organization for African Unity
ODA	Official Development Assistance
OECD	Organization for Economic and Cultural Development
PAMSCAD	Programme of Action to Mitigate the Social Costs of Adjustment
PDC	Provincial Development Committee
PDCs	People's Defence Committees
PHC	Primary Health Care
PNDC	Provisional National Defence Council
S&C	Savings and Credit Societies
SAPs	Structural Adjustment Programmes
SHG	Self-help Group
SPUs	Seed Production Units
TDC	Town Development Committee
UNCRD	United Nations Centre for Regional Development
UNECA	United Nations Economic Commission for Africa
UNICEF	United Nations Children's Fund
UNPAAERD	United Nations Programme of Action for African Economic Recovery and Development
UWT	Umoja wa Wanawake wa Tanzania
VIDCO	Village Development Committee
VSL	Vacuum Salts Limited
W/G	Women's Group
WADCO	Ward Development Committee

WB	World Bank
WCED	World Commission on Environment and Development
YSWGED	Yalta South Women's Group Enterprise Development

1

DEVELOPMENT FROM WITHIN?

The struggle to survive

Fiona Mackenzie

INTRODUCTION

Social struggle within rural Africa, whether it takes a visible form of co-ordinated action *vis à vis* local power structures or less overt but perhaps more sustained forms of 'everyday' resistance (Scott 1986) at intra or extra household levels, dispels any myth of an undifferentiated, immobile peasantry through whom a strategy either of 'grassroots initiatives and community self-management', as proposed by the ECA (Economic Commission for Africa) (UNECA 1989a: 47), or of greater community responsibility for 'development', in the words of the World Bank (WB 1989: 54), may be glibly promoted. Such strategies may indeed be a cheap 'development platform' (to paraphrase Watts, 1989: 6) at a political moment when the more difficult options of challenging the protectionism of industrial economies and renegotiating terms of trade are ignored and when state intervention is made the scapegoat for policy failure. But the proponents of these strategies, albeit they are analysing African crises from different ideological perspectives, fail to examine critically the outcome of their proposals. To call for the 'empowerment' of local people is to challenge social structure. Profoundly, one is dealing with 'politics' not 'policies', with 'struggle' and not 'strategy', as Weaver (1984: 138) points out.

In order to explore the interface between these antonyms, to investigate the nature of local action in the context of a deepening crisis in Africa, authors were invited to examine how members of communities in specific rural localities drew on their own resources, or attempted to gain access to resources controlled by others, as they sought to survive. A critical question which is raised by the authors' research is whether the action that they identify may be seen as being

1

a coping mechanism, as a result of the extent to which it meets the practical needs of everyday life, or as being of strategic significance (Molyneux 1985), in that it challenges social structure through a reconstruction of gender or of 'class'[1] and thus concerns empowerment.

The objective of this introductory chapter is to provide a conceptual context within which the case studies may be situated. As an entry point, the crisis in Africa is examined from the perspective of two multilateral institutions, the UNECA (United Nations Economic Commission for Africa) and the World Bank. Through an analysis of the contradictions posed, in particular by the Bank's agenda of continued structural adjustment, the opportunities for and constraints against local mobilization are exposed. Issues of sustainability and of popular participation are explored in the context of the debate between these two institutions.

Through the *African Alternative Framework to Structural Adjustment Programmes for Socio-economic Recovery and Transformation* (AAF–SAP) (UNECA 1989a), the ECA has brought to centre stage an alternative discourse for Africa to that of the World Bank, premised on human centredness. Reaffirming a commitment to an 'internally-generated self-reliant process of development' (Adedeji 1989: 5), which was first articulated in international fora through the Lagos Plan of Action and the Final Act of Lagos (1980), the Framework's rationale lies in 'immersing short term stabilization and adjustment with long term social and economic transformation' (UNECA 1989a: 33). Endorsed at the OAU summit in July 1989 by member states, AAF–SAP challenges blind adherence to the externally motivated prescriptions of stabilization and structural adjustment of the IMF and World Bank. Its progenitors argue, indeed, that these two institutions, in so far as their prescriptions divert attention from the root causes of crisis in Africa and focus instead on the 'symptoms and consequences of Africa's underdevelopment' (Adedeji 1989: 7), mask the nature of Africa's crisis and hence its sustainable resolution, thus contributing to its aggravation.

MEASURING THE ECONOMIC CRISIS

Methodologically, it is impossible to define either the crisis in Africa in general, or the impact of structural adjustment programmes in individual states specifically, on the basis of quantitative data (Berry 1984b: 61; Bienefeld 1989: 4–5; Loutfi 1989: 138–40; Watts 1989: 9).

Official statistics, or 'facts', as Michael Watts points out, are elastic. Frequently, the problem is less one of 'theory shaping fact' than of 'fiction masquerading as fact' (Watts 1989: 9).

Seldom are the contradictions that can be generated by reliance on macro economic data more flagrantly displayed than in the recent interchange of documents between the World Bank (WB/UNDP 1989) and the UNECA (1989b). While the WB/UNDP insist that SAPs have assisted economic performance and that strongly'[2] adjusting countries perform better than countries with 'weak' or no programmes (WB/UNDP 1989: 30, Table 20), the ECA accuses the Bank of faulty methodology. Selectivity in compilation, presentation and analysis of data, the ECA contends, has been exercised with respect to time periods under review, the establishment of base years, the inclusion or exclusion of data sets and the country groupings of African states. Groups of states have 'strong', 'weak' or no reform programmes, yet no definitions of terms are given and clusters are inconsistent with previous Bank practice (UNECA 1989b: 1-7). The data, the ECA proposes, 'seem to have been chosen, at least in some cases, to fit pre-conceived conclusions' (ibid: 6). The result is tautological; those countries that are 'winning' are those that are strongly reforming.

Where data are not available, optimism prevails, leading the Bank to conclude that 'the crisis seems less precipitous and the road to recovery more obvious and more manageable' (WB/UNDP 1989: 1):

> The growth that appears to be resulting, at least in part, from this reform and adjustment helps raise living standards overall and especially for the poor. The agricultural reforms that many countries have adopted, for example, increase the earning of small farmers – who make up about 80 per cent of the population of Sub-Saharan Africa and include most of the poorest people.
>
> (WB/UNDP 1989: iii)

Using the same data set, the ECA turns the table on the Bank, arguing that the latter's compilations for the period 1980–87 showed that countries with strong SAPs had the worst economic performance of any group.[3] The negative annual average growth rate of – 0.53 per cent of this group contrasts with + 2.00 per cent for countries with weak SAPs and a strong + 3.5 per cent for countries with no SAPs.[4] The negative growth rate of the strongly adjusting countries, the ECA points out, leads to an average growth rate for all SSA for

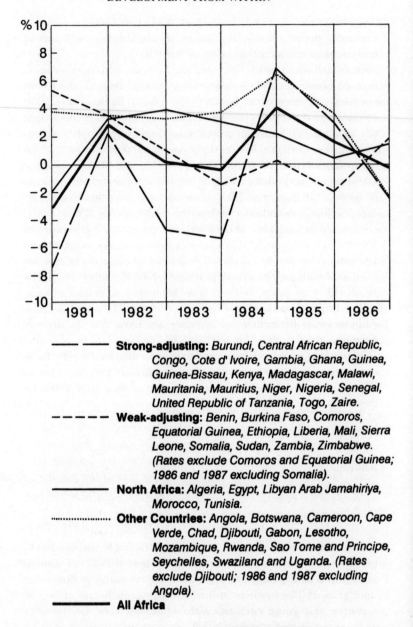

Strong-adjusting: *Burundi, Central African Republic, Congo, Cote d' Ivoire, Gambia, Ghana, Guinea, Guinea-Bissau, Kenya, Madagascar, Malawi, Mauritania, Mauritius, Niger, Nigeria, Senegal, United Republic of Tanzania, Togo, Zaire.*

Weak-adjusting: *Benin, Burkina Faso, Comoros, Equatorial Guinea, Ethiopia, Liberia, Mali, Sierra Leone, Somalia, Sudan, Zambia, Zimbabwe. (Rates exclude Comoros and Equatorial Guinea; 1986 and 1987 excluding Somalia).*

North Africa: *Algeria, Egypt, Libyan Arab Jamahiriya, Morocco, Tunisia.*

Other Countries: *Angola, Botswana, Cameroon, Cape Verde, Chad, Djibouti, Gabon, Lesotho, Mozambique, Rwanda, Sao Tome and Principe, Seychelles, Swaziland and Uganda. (Rates exclude Djibouti; 1986 and 1987 excluding Angola).*

All Africa

After UNECA 1989a: 22

Figure 1.1 Growth of gross domestic product in Africa[1]

4

the same period of + 0.24 per cent (UNECA 1989b: 10). Figure 1.1 illustrates the annual fluctuation for each of the groups, but it is important to recognize intra-group variation in economic perform-ance, which is dependent on a complex number of factors, including *inter alia* 'weather, conditions of commodity markets, inflow of external resources, the debt situation and structural determinants' (UNECA 1989b: 11). Thus, the ECA argues, 'any attempt to establish a one-to-one relationship between growth trends and the adoption or non-adoption of SAPs is prone to oversimplification and fallacy' (ibid.).

Equally alarming in the construction of an argument is the Bank's ability to use an indicator such as current account deficit (by its own reckoning illustrative of 'deterioration' of economic circumstance, WB 1988b: 2) as indicative of success (WB/UNDP 1989: 29). The Bank argues that only those countries with strong SAPs have been able to 'sustain a progressively widening current account deficit in 1986 and 1987' (ibid.: 28). For Africa as a whole, the deficit increased from $3.9 billion (US) in 1980 to $20.3 billion (US) in 1988 (Adedeji 1989: 10).

The implications of this situation for foreign-borrowing are clear. Although it is frequently pointed out that Africa's debt (which has risen over the decade 1978–88 from $48.3 billion [US] to $230 billion [US], Adedeji 1989: 10) is low, high debt service ratios cast into doubt the ability to sustain even current economic performance. In Ghana in November 1989, 70–75 per cent of export earnings were directed towards debt servicing (*West Africa*, November 1989). For Kenya, the equivalent figure during 1989 was close to 40 per cent (EIU 1989: 4). Even the Côte d'Ivoire, so long exhibited as a 'miracle' of economic performance, now recognizes the costs to political and economic stability of a soaring debt burden.[5]

The problems associated with the continued reliance on external sources for the direction of economies will be examined shortly but it is important to note here, in terms of methodology, that accurate data are more difficult to obtain now than a decade ago (Loutfi 1989: 142). This problem relates, in part, to cutbacks in government expenditure (ibid.) and, in part, to what is actually being counted. This concerns not only 'counted' crops such as cocoa in Ghana, which at various times has escaped official marketing channels by finding more lucrative markets in neighbouring Côte d'Ivoire or Togo (Bequele 1983; Green 1988) and so evading Ghanaian statistical records, but also the lack of visibility of a large sector of the economy, which Jamal (1988) refers to as the 'unconventional economy').

In the everyday struggle of survival in Africa, the majority of rural and urban households survive through juggling a myriad of activities. Even where there is access to a formal sector wage or salary, the reality for only a minuscule percentage of the households, this income provides but a small proportion of daily needs. At the height of the 1983 crisis in Ghana, brought to a head by the drought and rapid inflation of food prices, UNICEF estimated that the minimum wage would cover 2.6 per cent of a minimum 'socially acceptable household budget' for a family of five; the salary of a middle level civil servant would cover 5.9 per cent (Loxley 1988: 9). Households of five members in the low income areas of Accra, Nima and Maamibi spent over eight times the minimum wage in order to survive (ibid.).

Green's (1988: 15) reconstruction of the budget of a Ghanaian household with a male clerk as wage earner in late 1985 illustrates where, in an urban area, the extra income is generated. Wages here provided 25 per cent of household expenditure. A further 25 per cent came from housing and other allowances, tips and sales memos. The wife's income from the sale of vegetables and other 'informal' sector activities and income from other household members accounted for the remaining 50 per cent. With a freezing of wages in 1988, the relative proportion of household income from this source will have decreased, with a return closer to the figures of 1983.

Phenomenal growth in the 'unconventional' or 'informal', but largely uncounted, sector is widely reported as a response to economic crisis in Africa (for example, Colclough 1988; Evans 1989; Green 1988; Loxley 1988; Maya 1988; Onimode 1989a; Tibaijuka 1988). Individuals with or without wage or salaried employment frequently have three or more 'jobs'. The growth in petty trading among women and children has been particularly rapid, yet concomitant with a squeezing of retail margins. Loxley (1988: 39) suggests that, in contrast to wholesale traders whose monopolistic organization allows them to compensate for measures of structural adjustment, such as devaluation of the currency and import liberalization in Ghana, retailers are faced with increasingly intense competition, leading frequently to depressed earnings from this source.

For measuring the economic crisis, there is a woeful lack of adequate data on the distributional consequences of current structural adjustment policies. The political consequences of this omission became most obvious in the riots of December 1986 and June–July 1990 in copperbelt towns in Zambia. Extensive rioting followed the

6

removal of the maize–meal subsidy by the government (Colclough 1988: 60; Young 1988: 25; *Globe and Mail*, 6 July 1990) in a context where miners, particularly those in the middle and top grades, were disproportionately affected by the economic reform programme. Loxley (n.d: 25) illustrates how, for these two groups of miners, real wages fell 77 and 84 per cent respectively between 1981 and 1986. In contrast, the wages for lower paid miners fell 56 per cent in real terms. Both real per capita GDP and real value-added in the mining industry dropped by 19 per cent over the same period.

Available data from critical sources draw attention elsewhere to the widening disparities among social classes as a result of current policy implementation (Bienefeld 1989; Colclough 1988; Longhurst *et al.* 1988; Loxley 1988). Bienefeld (1989: 70–2), for example, illustrates how, in Tanzania, social polarization is growing rapidly as a consequence of an Economic Recovery Programme which is squeezing the majority of the peasantry who were previously 'protected' by policies which accorded priority to food security. The abandonment of pan territorial pricing exemplifies one instrument operating here. However, recent studies commissioned to support the work of the Commonwealth Expert Group on Women and Structural Adjustment, published by the Commonwealth Secretariat (1989), indicate that analysis of the distributional impact of economic reform programmes must proceed beyond an investigation of urban/rural and rich/poor to an examination of the differential impact on individual household members.

The conceptual point that analysis, or measurement, of the economic crisis must focus on the intersection of class and gender may be underlined through an examination of the repercussions of removing subsidies of food staples, such as maize, in urban areas and raising producer prices of the same crops in rural areas. Evidence from Munachonga's (1986) survey, Sanyal's (1984) doctoral thesis on urban agriculture, and the study by Evans and Young (1987) of household labour allocation in maize producing households in Northern Province, Zambia, which is drawn on by Evans (1989), provides some insights into this issue.

The maize subsidy, Evans (1989: 28) notes, was the 'cornerstone of Zambia's pricing policy', since maize consumption provided 50 per cent of calorie intake for the population. Removal of the subsidy resulted immediately in a substantial increase in the cost of living. In urban areas, where the lowest income groups spent 77 per cent of their monthly income on food, the impact was dramatic (Evans 1989:

7

33). Munachonga (1986) argues that the strain created particular hardships for women on account of intrahousehold practices of income allocation and control. Her data indicate that men's differential access to income meant that women seldom controlled sufficient resources to provide meals for the family. Women's responses, in addition to borrowing money, were to change consumption patterns through buying less food, buying cheaper and especially less protein-rich food, and reducing the number of meals per day. More positively, some women in Munachonga's survey began to grow their own food. Among poorer households, this frequently entailed travelling to vacant land in the urban periphery; for less poor households, plots close by were common. Production in both cases was primarily for household consumption. From Sanyal's (1984) study, it is evident that vegetable growing could provide for half of household expenditure on food. For 'better off' low income households, 13 per cent of expenditure on food originated from home cultivation (Evans 1989: 31). Female-headed households were particularly vulnerable to food price increases, since not only was their labour insufficient to engage in urban agriculture but also there were fewer economically active members on whom to draw for resources (Evans 1989: 34).

The differential impact on women, contingent on socioeconomic and marital status, of the elimination of the maize subsidy is further evident in rural Zambia: the study by Evans and Young (1987) in Northern Province distinguishes among households on the basis of whether or not maize is produced for the market. Evans (1989: 37–8) points out that only a small percentage of small-scale producers benefited immediately from an increase in the producer price of maize, as the majority of farmers grew cassava, millets and sorghum, the prices of which were not increased. Drawing on an ILO report of 1987, Evans points out that average per capita cash income for the economically active population in this sector actually dropped by 13 per cent between 1976 and 1985. 'Rural producers reliant on selling "traditional" food crops (predominately women)', she writes, 'found their incomes being squeezed by stagnating producer prices and rising transport costs on the one hand, and rising consumer prices on the other' (1989: 39). Frequently, in households headed by women, whether widowed, separated or in a polygamous marriage, insufficient labour and lack of capital with which to purchase seed and fertilizer meant that maize was not grown (Evans 1989: 41).

Among those households which produced maize, the study by

Evans and Young (1987) found evidence that an increased focus on production for the market intensified gender struggle within the household. The labour allocation survey revealed 'that a major interest amongst men in cultivating hybrid maize was to gain direct control over cash income' (Evans 1989: 42). Despite the fact that women provided more labour time in maize production than men, men managed to gain access to the proceeds of the crop through payments from the marketing union and through controlling expenditure. Evans notes that while some women were given money for clothing for the household this did not cover other expenses. Although the survey indicated that women retained 'more or less direct control' of income from other crops, the increased hectarage under maize meant that less land was allocated to millets, beans and groundnuts, and that women had less time to spend on these crops. Also, prices for these crops were much lower than for maize. Further, as Evans notes, 'the purchasing power of such income was declining as crop prices failed to keep pace with the prices of processed and manufactured goods' (ibid.). Evidence from Ghana and the Côte d'Ivoire similarly illustrates the lack of benefit accruing to women from incentives for the production of cash or export crops (Common-wealth Expert Group, 1989). Women's withdrawal of labour from export crop production as a consequence of lack of control over the proceeds is well documented (Carney 1988; Mackenzie 1986; Mbilinyi 1988).

The only exception to this trend of declining income for female farmers, as explored in the Zambian case study, is that of women brewers of beer which is made from millets and maize. This most lucrative form of non-farm income had two negative side-effects, however. First, it led to a decline in food stocks available for house-hold consumption. Second, an increase in the incidence of domestic violence is attributed to growing alcoholism and to women's increased income which threatened the intrahousehold balance of power (Evans 1989: 43).

A negative change in women's bargaining power *vis-à-vis* men is seen to result from other policies carried out under economic reform programmes. Evidence from Tanzania (Tibaijuka 1988), Nigeria (Onimode 1989a) and Zambia (Evans 1989) illustrates the impact on women of cutbacks in formal sector employment. Although Evans (1989: 21) points out that, in Zambia, cutbacks in the formal/public sector had a proportionately greater impact on men because more men are employed in this sector, retrenchment has affected chiefly low-skilled, low wage labour, and so women have been heavily affected. This fact, and the inadequacy of the wage or salary to ensure household reproduction, has led to an explosion of activity in the 'informal' sector, as has already been identified. Much of the increased activity comes from a multiplication of women's labour in this sector. Tibaijuka's (1988) survey of 134 women in low income areas in Dar es Salaam is instructive here. All of the women were involved in income-generating activities, such as petty trade, beer brewing, poultry raising, tuition and tailoring. Seven only of the 134 women had 'professional' jobs. Despite this active and increased participation in economic activity, 58 per cent of the women interviewed said that they had been forced to reduce the number of meals from three to two in their households; 61 per cent indicated a reduction in protein-rich foods (1988: 37–40).

Clearly, as has been pointed out previously, intense competition in the 'informal' sector, as more and more people strive to survive, squeezes profit margins in a context where declining real incomes decrease the purchasing power of urban consumers for goods and services from this sector (Evans 1989: 19). Women are particularly affected. They are likely to be 'downwardly mobile' because of involvement in less profitable enterprises compared to men, and a lack of access to capital and to material resources, for example, premises for storage (Evans 1989: 21).

Untangling the distributional effects of SAPs is complex and requires much further research which links micro analysis with an understanding of macroeconomics at the national and international levels. However, as has been demonstrated through discussion of research which was conducted under the auspices of the Commonwealth Secretariat, data are sufficient to counter arguments such as Loutfi's (1989: 139) that, as the 'informal' sector is a source of economic refuge, the policy impact of SAPs is of little relevance to the functioning of the economy. An intensification of social struggle, as polarization of wealth becomes more deeply embedded, is played out at the intrahousehold level as individuals strive to ensure household survival. The 'African social fabric', as AAF–SAP contends, is in danger of collapse (UNECA 1989a). It is in this light that Adedeji (1989: 12–13) questions whether the SAPs are even a 'temporary palliative' for the problems that Africa faces. They have led, he argues, to an 'erosion of national sovereignty', in terms not only of a hijacking of economic policy to external design but also of a neglect of the human dimension of the crisis.

THE 'AFRICAN SOCIAL FABRIC'

In part, the neglect to which Adedeji refers is reflected in a reversal in the 1980s of gains which were made in the previous two decades in levels of nutrition, health, the provision of safe water and education. Green (1988: 7) cautions that there is a danger, when analysing the human condition today, of ascribing the prevalent decay solely and universally to SAPs. In Ghana, he notes, only belatedly did the Provisional National Defence Council (PNDC) begin to address the deteriorating social conditions that preceded the crisis of 1983 and agreements with the World Bank and IMF, by means of the Programme of Action to Mitigate the Social Costs of Adjustment (PAMSCAD) (Ghana 1987). The central issue in analysing this crisis, fundamentally of social reproduction, concerns the extent to which present policies not only undermine previous gains, the focus of Green's (1988) analysis, but also shape, through an intensification of class and gender contradiction, the ability of people to actualize their own survival. In this section, the distributional effects of declining public sector support for health and education are discussed. Again, as would be expected, and as argued so vehemently in UNICEF studies, the cost is disproportionately borne by the poor, by women and by children.

Table 1.1 illustrates the contrasting situation for Ghana and Zimbabwe, countries for which comprehensive data are available from UNICEF's case studies. The dramatic decline in wellbeing which is illustrated in Ghana is related to economic crisis and cutbacks in government social expenditure, both before and after agreements with the IMF/World Bank (Table 1.2). Cuts in Ghana in public expenditure for health are reflected in decreased access to and decreased quality of these resources, in contrast with Zimbabwe. Maya (1988: 27) notes that in Zimbabwe government spending since 1983/4 through Labour and Social Welfare, the main channel for public sector social spending, has decreased by 50 per cent, but that this figure is probably misleading, reflecting temporary payment increases in the face of drought and demobilization. Maya argues that health in particular has remained a government priority even in the face of policies of structural adjustment.

In Ghana, cutbacks have been massive. With respect to health personnel, Ghana lost over half of its doctors between 1981 and April 1984 (Commonwealth Expert Group [CEG] 1989: 57). In 1983, only one of the graduates of the University of Ghana Medical School,

11

Table 1.1 Ghana and Zimbabwe: summary of UNICEF's process and outcome indicators

	Ghana (1979–85)		Zimbabwe (1980–84)
	late 1970s	*1980s*	*1980s*
Process indicators			
Average calorie availability as % of requirements	88%	68%	N/A
Access to/use of health services	Decrease of 11% p.a.		Increased coverage of health care and preventive service
Access to/use of education services	Positive change (Primary enrolment rate [6–14 years] constant)		Postive change (Primary enrolment rate [6–14 years] up to 20% per annum)
Outcome indicators			
Infant mortality rate[1]	86%	107%	Positive (qualitative assessment)
Child death rate[2]	15%	25–30%	N/A
Pre-school malnutrition	35% (1980)	54% (1984)	Constant (1982) 18–22% (1984) 16–25%
Primary school quality	Negative		N/A
Disease prevalence	High morbidity, increased yaws and yellow fever		Positive decline in incidence of disease
Population below poverty line: rural urban	60–65% (1974) 30–35% (1974)	70–75% (1984) 45–50% (1984)	

Source: Cornia 1987: 24–6

1 IMR, per 1,000 live births
2 CDR, per 1,000 of the population aged 0–15 years.

12

Table 1.2 Sub-Saharan Africa: annual percentage cuts in per capita GDP
and health and education expenditure
(cumulative totals given in parentheses)

	Health 1979–83	Education 1978–83	GDP 1980–5
Ghana	– 15.8[1] (– 40.3)	– 9.5[1] (– 25.9)	– 4.4 (– 20.1)
Malawi	– 9.8[1] (– 26.6)	+ 7.0[1] (+ 31.1)	– 1.1 (– 4.9)
Sudan	– 9.5 (– 32.9)	– 16.8[1] (– 42.4)	– 2.6 (– 12.3)
Togo	– 7.5 (– 26.8)	+ 3.3 (+ 13.9)	– 3.7 (– 17.2)
Liberia	– 6.9 (– 24.9)	– 0.6 (– 2.4)	– 7.1 (– 30.8)

Source: Compiled from Table 3.3, Pinstrup-Andersen *et al.* 1987: 76

1 1979–82.

Korle Bu, remained in the country (personal interview with Dr
Ampofo, Professor of Medicine, 1986). One-twelfth of Ghana's
nurses left in 1982 (CEG 1989: 57). Pinstrup-Andersen *et al.* (1988:
72) note that cutbacks in financing health in Ghana at the height of
the economic crisis in 1982–3 resulted in inadequate infrastructure
for the basic health programme. For example, in the Accra region, 20
of the 57 refrigerators essential for maintaining a cold chain for
vaccines were unusable. The outcome indicators (Table 1.1) of child
survival and welfare reflect this situation.

Evidence from Zambia parallels Ghana's experience, and has been
directly attributed by the Bank to the reform programme carried out
under its auspices (Evans 1989: 22). Again, a decline in the provision
of health care is evidenced by a fall in real per capita expenditure; a
loss of medical personnel (between 1983 and 1986, the ratio of doctors
per 100,000 people fell from 13 to 5); and a drop in the real value of
the drugs budget (in 1986 under 25 per cent of its 1983 value) (ibid.).
The provision of health care in rural areas has particularly been
jeopardized. In a 1986 survey, 73 per cent of rural health centres had
unfilled staff positions, 51 per cent of the vehicles were in disuse, and
there was a lack of essential drugs and equipment. (Drugs were available on the parallel market but at a highly inflated price; Evans 1989:
44.) As Evans points out, those services most used by women and
children required increased costs, both directly for increased distances, and indirectly as more time away from productive work
ensued (ibid.). Where user fees are required, as in urban centres in
Zambia (Evans 1989) or more widely spread in Nigeria (Onimode
1989a), data indicate decreased use of health facilities by the poor.

13

The long term effects of these policies may not be assessed at present but an increase in maternal and child mortality, particularly in the perinatal period, and an increase in low birth weights are seen by researchers in Tanzania (Tibaijuka 1988: 12) and Zambia (Evans 1989: 24) as evidence of an ominous trend. The escalating decline in adequate nutrition supports this view. Data in Table 1.1 illustrate the situation for Ghana. As in Zambia, malnutrition is associated with the dietary changes documented in the previous section, which are associated with declines in real income for the majority of the population in Ghana and Nigeria (Onimode 1989a: 15). It is also associated with a change in balance of power within the household. Evans (1989: 45) draws on a 1986 survey in Zambia to argue that 'child welfare in households with a variety of individual farming and food processing enterprises is significantly better off than in households whose resources and labour are concentrated in the hybrid-maize farming of one member'. Evidence from Western Kenya (Kennedy and Cogill 1987) and Longhurst's (1988) analysis of the relationship between export/cash cropping and intrahousehold malnutrition suggests that the degree to which a woman controls the income from such crops is a critical determinant of nutritional adequacy among children. In Zambia, Evans (1989: 24) draws attention to the increase in hospital deaths of children which are 'attributed' to malnutrition; they are a minute percentage of those suffering from inadequate nutrition.

Cutbacks in government budgets for health inevitably shift the burden of health care to the household, causing acute problems for low income households, and particularly women, as the prime individuals responsible for such care. The opportunity costs in terms of income generation or agricultural production, and thus potential ability to increase the level of consumption, are high. As the CEG summarizes, the strain on women's labour time increases:

> More time as producers, working outside the home, to contain the falls in household income; more time as home managers and mothers to protect family health and nutrition in the face of rising food prices, reduced incomes and reduced social services; and more time as community organisers to help counter some of the adverse effects through community action.
>
> (CEG 1989: 62)

With respect to education, cutbacks in expenditure and the implementation of user fees again affect the poor and women differentially. In Zambia, where user fees were not as yet introduced for primary or

14

secondary education in 1985, it was estimated that parental expenditure for items (uniforms, books, paper, transport) required for one child to attend primary school averaged one-fifth of per capita income (CEG 1989: 56). For secondary schooling, the cost was nearly half of the average per capita income (Evans 1989: 27). Given the existing bias against female education, which is evidenced in literacy rates for 1985 for the country as a whole (48 per cent of the female adult population, compared to 29.5 per cent of the male, was illiterate), higher female dropout rates and lower participation rates in secondary and tertiary educational institutions, as Evans (1989: 26–7) argues, discrimination against women is likely to be 'aggravated' during a period of structural adjustment.

Direct evidence of the impact of user fees on school attendance comes from Bendel State, Nigeria, but unfortunately this is not disaggregated by sex. Subsequent to the introduction of school fees in 1982, primary school enrolment fell from 90 to 60 per cent within eighteen months (Cornia 1987: 34). An escalation of these trends may be anticipated, with a disproportionate effect on women/girls from low income households, as it is their unpaid labour time which is bearing the hidden cost of cutbacks in public social sector spending and their participation in education which is curtailed first under conditions of economic stress. An intensification of female labour as a result of economic reform programmes is widely documented (CEG 1989; Evans 1989; Tibaijuka 1988).

ENVIRONMENTAL COSTS

The relationship between the economy and the environment has until recently been invisible in analyses of the economic crisis in Africa. However, inspired no doubt by the popular attention given to the Brundtland Report (WCED 1987), both AAF–SAP (UNECA 1989a) and the Bank's publication *Sub-Saharan Africa: from crisis to sustainable growth* (WB 1989) include the environment as a cost connected to economic insecurity.

Although given only brief attention, AAF–SAP (UNECA 1989a: 11) links the aggravation of an environmental crisis to poverty and the increasing severity of the economic crisis: '[I]ncreased poverty induces the poor to degrade further the physical environment as they struggle to eke out a precarious living'. While AAF–SAP may be criticized for not integrating the relationship between poverty and environmental degradation more centrally in its analysis and recommendations, its statements are less schizophrenic than those of the Bank.

Early in its report (WB 1989: 44), the Bank characteristically holds up population pressure as the primary cause of environmental degradation. Subsequently, the link between poverty and environmental cost is made, albeit within the context of a neoclassical growth paradigm:

> low prices for produce coupled with uncertain land tenure make conservation financially unattractive. . . . A more prosperous rural environment – resulting from the removal of price, exchange rate, and fiscal and other distortions together with greater security of tenure and improvements in productivity – is necessary for farmers, forest dwellers, and pastoralists to have an interest in conservation. Conservation will fail unless it appeals to the farmer's self-interest.
>
> (WB 1989: 101)

The argument that policies pushed by the Bank, specifically those geared to an expansion of export production and hence further integration into the world market system, increase the contradiction between the economy and the environment is ignored.

But it is precisely through what Redclift (1987) has termed the 'internationalisation of the environment' that the contradictions for 'sustainable development' are exaggerated. The central contradiction with advanced capitalism, in terms of its relations with the third world, concerns the way in which 'the labour process, the means by which the social mediation of nature is achieved, succeeds in transforming the environment in ways that ultimately make it less productive' (Redclift 1987: 51).

Redclift's argument is succinctly summarized at the conclusion of his earlier book:

> poor people impose excessive strains on the carrying capacity of the natural environment because of the structural demands imposed on them. The need to increase cash income, repay debts and meet the necessities of the household impinge upon poor people while they are held in a vice by the terms of trade which govern intersectoral and international relations. As 'development' removes them from control over their own environment, this control is assumed by transnational companies and capital-intensive technologies. As some activities,

16

especially those of women, are transferred from the household to the market place, the environment is relocated not as part of a *local* system of production, but as a link in the international division of labour.

(Redclift 1984: 130)

The solutions, to follow Redclift's (1987) reasoning, lie not with technological tinkering ('environmental management' carries with it a 'facade of technical objectivity'; 1987: 138) but with a profound redistribution of the means and resources of production. In this light, Redclift (1984: 23) argues, it is 'intellectually dishonest' to suggest population increase as a cause of environmental deterioration.

Redclift does not pursue his analysis of the political economy of the environment 'downwards', as Blaikie (1989: 26) suggests, to examine the 'dialectic between environmental and social change' (ibid: 22) in terms of differential gender access to and control over land, livestock, labour, and decision-making at the level of the household. Yet it is at this level, it may be argued, that the 'simple reproduction squeeze' (Bernstein 1979: 427) is most keenly felt. In Africa south of the Sahara, where women are the prime agriculturalists, maximization of short term economic gain through mining of the resource base may outweigh the long term consequences of this action. The option value (Blaikie 1989: 22) of investing in the maintenance of soil fertility, for example, may decrease as a consequence of the stress to meet immediate needs. It may become a less viable strategy for women where security of tenure is threatened through the increasing polarization in rural society and where they need to strengthen social relations which ensure access to land, to follow Berry's (1989: 44-8) argument, demands the diversion of surplus from direct investment in production. As Berry (1989: 49) argues, where there is substantial insecurity, common survival strategies include 'maintaining as liquid a position as possible in one's asset holdings. People are reluctant to tie up capital or fixed assets or long-term projects and prefer to hold commodities . . . or to specialise in activities with rapid turnover (such as trade) rather than in those with long gestation periods.'

As relations of production change within the household under economic stress and as commoditization of production increases, the household becomes a more 'deeply contested terrain' (Watts 1989: 12; Guyer and Peters 1987; Carney 1988; also Vaughan 1985). One outcome, as women strive to meet their responsibilities for household maintenance, is indeed to increase pressure on the land (Mackenzie

17

1991). As undermining of the resource base continues, a cycle of environmental decay increases rapidly the reproduction costs of the household (Blaikie 1989: 28). Thus the contradiction between land use management for agricultural production and the long term sustainability of the resource base is intensified.

As a corollary to this argument, the integrity of local environmental knowledge systems (Belshaw 1979; Kjekshus 1977; Igbozurike 1971; Richards 1979, 1983, 1985) is not in question. Use or misuse of the environment by individuals from specific socioeconomic strata is, as Watts (1989: 15) explains, 'a function of the intersection of resource managers with extrahousehold, non-local circuits of accumulation and surplus extraction'. Nevertheless, it may be argued that, as the option value of maintaining a sustainable resource base declines and as cropping strategies based on exportable commodities prevail, a further subjugation of the 'ecological specificity' (Richards 1983: 56) of local knowledge is likely. Women's relationship to the land alters through a reproduction squeeze, commodity production is maximized at the expense of long term land sustainability, and thus a particular knowledge base is threatened. The ecological costs of the devaluation of such knowledge are intimately linked, as Stamp (1989: 129) illustrates, with the pervasive imposition of Western 'knowledge' on Africa and the construction of knowledge about Africa through a process whereby dominant power structures are reinforced. It is with this process, yet a further implication of Africa's loss of sovereignty, that the recent ECA documents take issue on a large scale.

COUNTERING THE EROSION OF SOVEREIGNTY

Diagnosing the crisis in Africa in terms of external dependence, which created social and political structures leading to a polarized society, the Lagos Plan of Action (LPA) and the Final Act of Lagos (1980) sought resolution of the crisis through a long term strategy of self-sustaining development based on the maximum internal use of Africa's resources (UNECA 1980). Faced with a balkanized continent, the LPA placed emphasis on national and collective self-reliance, the alleviation of poverty and a raising of living standards, and an egalitarian society (Adedeji 1985; Asante 1985; Lardner 1985). In this analysis the environmental costs of the prevalent approach were ignored, although the ecological implications of policies towards agriculture were considered in commentaries (Daddieh 1985).

However, by 1985, with a deepening economic crisis, the analysis in the African Priority Programme for Economic Recovery 1986–90 (APPER) had been modified, despite an affirmation of the principles of the LPA, to include a frank acknowledgement of 'internal factors' which contributed to the crisis. The lack of implementation of the LPA was considered to be due to 'insufficient structural transformation and economic diversification' from an inherited colonial structure (UNECA 1986: 7). As the crisis intensified, policy makers shifted their attention to short term crisis management through stabilization and adjustment programmes (UNECA 1989a: 10).

Identification of endogenous factors in APPER indicated considerable responsibility within Africa for the crisis. Among the factors were: structural imbalance (e.g. urban–rural; disparities in income distribution); population increase; the inadequacy and misdirection of human and financial resources; inappropriate economic strategies and policies; poor economic management; inadequate institutional and physical infrastructure; the 'persistence of social values, attitudes and practices not always conducive to development'; and political instability (UNECA 1986: 9). Among the exogenous factors identified were: international recession; commodity price collapse; adverse terms of trade; decline in real terms of ODA; increasing protectionism on the part of northern markets; high interest rates; currency fluctuations; and high debt and debt servicing obligations (ibid.: 10).

As a result of this analysis, solutions to the crisis were sought through a balance between external and internal measures. APPER refers to 'shared responsibilities' and 'a genuine partnership' between Africa and the international community (UNECA 1986), a co-responsibility which was accepted in the UN Programme of Action for African Economic Recovery and Development 1986–90 (UN-PAAERD) (UN 1986a). The approach was expected to lay the basis for self-sustained development but subsequent years have proved this not to be the case, as illustrated in earlier sections of this chapter. An ECA survey of mid-1987 documented not only the extent to which African countries were adopting policy reforms under SAPs, but also the lack of support from the international community. Ironically, the year 1986 saw a dramatic drop in the terms of trade of some of Africa's principal exports (Adedeji 1987).

An intensification of the crisis by the end of the 1980s led to a re-examination of previous analysis and policy recommendations on the part of the ECA. Remarkable in the vehemence of the language used, AAF–SAP (UNECA 1989a) considers that the growing reliance on the direction of African economies by multilateral institutions (the IMF and World Bank), chiefly through the implementation of SAPs, contributes to the crisis rather than resolves it. SAPs, it is argued, deal with symptoms of the crisis, such as budgetary and external disequilibria, rather than root causes; the human costs of SAPs are unsustainable and undermine long term, sustainable development. AAF–SAP notes:

> Unless there is an immediate amelioration in the conditions of the vast majority of the African population, there is a real danger of a systemic breakdown in the socio-economic fabric and the supporting natural environment.
>
> (UNECA 1989a: 10)

Regaining control over the direction of African economies and human resource development is identified as a key to the future:

> The gradual erosion of sovereignty implied in the growing role of officials of international financial and development institutions and donor agencies in policy design, implementation and monitoring without any accountability to the people of Africa will be reversed. . . .
>
> (UNECA 1989a: 49)

AAF–SAP is a 'blueprint' providing the 'broad guidelines and strategies' for long-term, 'human-centred' and 'holistic' development to be achieved through the 'immersion' of short term adjustment with long term social and economic transformation (UNECA 1989a: 33). AAF–SAP argues that

> Africa has to adjust. But, in adjusting, it is the transformation of the structures that fundamentally serve to aggravate the African socio-economic situation that constitutes the focus of attention. As such, adjustment and transformation must be conceived and implemented as inextricably linked and intertwined processes such that progress will be made simultaneously on the two fronts.
>
> (UNECA 1989a: 33)

The dichotomy between long and short term is resolved, in the document, through the 'endogenisation of development' (UNECA 1989a: 13).

Analysis of Africa's predicament proceeds, according to AAF–SAP, from a political economy perspective. In fact, this includes neither a debate over the role of the state nor a serious attempt to break

20

out of the duality (external/internal) of the conceptualization of previous documents. There is, however, a significant broadening of the field of discussion.

Consistent with previous theorizing, AAF–SAP again challenges 'the outward orientation' of consumption and production, arguing for the transformation of economic and social structure at the level of production (UNECA 1989a: 12). Principally, this will entail breaking the 'apron-strings of structural and relational dependence' which is most visible in the production of a few primary commodities for export. Africa will not make 'a development break-through' by following this route, it is argued: trade 'dependence' must be changed to trade 'viability' (UNECA 1989a: 13) and this is to be achieved primarily through an increase in intra-African trade. Only through this route, it is suggested, can there be a curtailment of structural dependency, the outward manifestations of which include a 'narrow, disarticulate production base with ill-adapted technology' (UNECA 1989a: 2), a neglected informal sector, a rising debt burden, capital flight, the loss of skilled personnel, declining institutional capability and deteriorating terms of trade. The environment and its degradation are included as an integral component in economic analysis (e.g. UNECA 1989a: 4, 11).

In contrast to earlier analysis, socio-political structure achieves greater visibility in AAF–SAP. The document points to the increasing social differentiation in African society – between classes, gender and ethnic groups – on the basis of economic and political cleavage, and the danger of imminent breakdown. But it also draws attention to the lack of 'democratic political structure' (UNECA 1989a: 7) within African society and calls for democratization of society and accountability of those in power to their electorate.

On a broad scale, AAF–SAP tackles the relevance of present institutional capability in resolving the crisis. Public administration is charged with a 'siege-mentality' which is responsive to '*ad hoc* crisis management' (UNECA 1989a: 33) at the expense of long term economic planning. Blame is placed on the increasing role of 'foreign experts and managers' and on the continuous rounds of negotiations with aid agencies and with creditors for debt rescheduling. The result is that 'the scope for independent policy-making and rational economic management in Africa has gradually diminished and narrowed' (UNECA 1989a: 8). Reinforcing national economic management capability is a key to restructuring.

Policy instruments which are designed to address these questions are organized in the Framework broadly around four themes: the strengthening and diversification of production capacity, including a focus on agricultural investment for food production; improving the level of income and the pattern of distribution; redirecting public expenditure towards measures to promote the satisfaction of basic needs (limiting debt service ratios is one instrument through which this may be achieved); and fostering appropriate institutional support to ensure adjustment with transformation (UNECA 1989a: 38–43). It is in this last context that community institutions, specifically indigenous NGOs and self-help programmes, are identified. The authors of AAF–SAP write as follows:

> At the institutional level, there is need for excessively central-ised bureaucracies to yield to local decentralisation, grass-roots initiatives and community self-management. . . . [T]he increas-ing role of the people in adjustment with transformation should facilitate the functioning of a system of checks and balances and safe-guard against bureaucratic excesses.
>
> (UNECA 1989a: 47)

Further, AAF–SAP's implementation

> must be based on a genuine and active partnership between the government and people through their various political, social and economic organisations at national, local and grass-roots levels.
>
> (UNECA 1989a: 49)

'Vigorous' government support for grass-roots' initiatives is called for (ibid.).

The extent to which such grass-roots initiatives may facilitate development 'from within', i.e. may be strategic in intent rather than merely coping with crisis, is the central question of this book. It is dis-cussed in this chapter after the Bank's proposals are analysed. It is important to note at this point, however, that AAF–SAP fails to consider the socioeconomic climate, the degree of differentiation, within which community initiatives are to flourish. 'Differentiation' may appear in the Framework's analysis, yet a fault line between theory and prescription emerges, owing to the failure to investigate social relations which have led to present practice. In AAF–SAP, as in previous UNECA documents, an underlying contradiction is buried in the supposed neutrality of bureaucratic discourse; namely, that those accepting the report were themselves assisted in reaching their position by market forces, particularly through integration into the international economy and present 'development' approaches, and yet are to be the instruments to foster 'another' development.

Community initiative is one of several recurring themes which are identified as having significance in terms of ways forward in AAF–SAP but which are inserted by the World Bank, through its recent response to the ECA, *Sub-Saharan Africa: from crisis to sustainable growth* (WB 1989), into an ostensibly alternative analysis. Acknowledging here the depth of the crisis in Africa, that social progress is in danger of erosion as hunger increases and as ecological degradation accelerates, the document similarly argues that the approach to development must be broadened from purely monetary considerations to become human centred:

> Improving health, expanding education, ensuring food security, creating jobs: these are the priorities shared by all the partners in Africa's development.
>
> (WB 1989: 186)

Continuing, the Bank argues that 'deep transformation of the production structure is required' through a long term strategy (ibid.: 186). Successful SAPs, the fulcrum of sustainable change, must go beyond stabilization in order to 'achieve a genuine transformation of production structures' (ibid.: 189), taking account of the social impact of adjustment measures. There is to be more investment in education, health, science and technology, infrastructure and environmental protection.

In the search for common 'high ground' in the discourse on African crises, the Bank also talks of equitable growth, not in terms of a 'distribution of wealth' but concerning 'access to assets and poverty alleviation' (WB 1989: 38). Self-reliance is acknowledged as a legitimate notion, albeit framed within arguments of comparative advantage. An increase in intra-African trade is seen as one route to this end. A continued insistence on a retraction of state intervention is tempered by an acknowledgement that short term intervention, for example in the area of food subsidies or in the provision of some public goods, may be necessary (ibid.: 55)

However, such calls, as well as those for greater democratization, consonant at the level of rhetoric of those of AAF–SAP, lie uneasily with the renewed commitment to the free play of market forces. As Hutchful (1990) argues, schizophrenia is evident in philosophical and practical posturing; in the tendency to mask old policies with new slogans; and in discussions which deal simplistically with some very difficult issues, such as those of local empowerment.

Consistent with previous analysis, namely the 'Agenda for Action', or the Berg Report as it is frequently named (WB 1981), and *Toward sustained development in Sub-Saharan Africa* (WB 1984a), the present report dwells on internal explanatory factors. A brief consideration of terms of trade, for instance, admits that for poorer sub-Saharan countries the situation is particularly difficult but that southern Asian countries in similar circumstances have 'coped better'. Africa's deteriorating economies, the Bank reasons, 'must be attributed in large part to a combination of high population growth and low GDP growth (WB 1989: 25). Population, as in previous reports, in relation to 'development' and, incidentally, environmental degradation (ibid.: 22), becomes part of a simple Malthusian equation, dissociated from historical context. Declining returns on investment are blamed for Africa's increasing uncompetitiveness internationally in agriculture and industry; in turn, poor public sector resource management and 'bad policies' are viewed as culprits. In the words of the Bank: 'Together these have undermined the efficiency of the private sector and have added greatly to the high cost of doing business in Africa' (WB 1989: 27).

Despite a partial apologia for co-responsibility for 'inappropriate' investment (WB 1989: 27), the Bank once again casts public sector management as the scapegoat for economic failure, and so continues to rely on policy instruments drawn from classical economic theory. In the words of AAF–SAP (UNECA 1989a: 18): 'output, employment, and prices (including wages, interest rates and the exchange rates) are best determined by the free play of market forces, and . . . prices are the most effective instruments for the efficient allocation of resources'. The main components of this 'monetarist jigsaw' (Adedeji 1989: 16), whose composite picture varies little from country to country, include: demand restraint, primarily through cuts in government expenditure, public sector employment and real wages; a pricing policy which aims at decontrol of price structures, including the elimination of subsidies and the imposition of user fees for public services; trade liberalization involving currency devaluation and foreign exchange auctions; credit reform principally through the setting of higher interest rates and the provision of increased agricultural credit; privatization of, for example, parastatals; and institutional 'strengthening' (Commonwealth Expert Group 1989: 45; Loxley 1988: 12–17).

It is against this continuation of previous economic direction, of orthodox structural adjustment, that the Bank's emphasis on community initiative may be assessed. Is local level empowerment part of the 'arsenal of new conservatism' emanating from the US (Hutchful 1990), the cheap 'development platform' of Watts (1989: 6)? Or is it, indeed, a sustainable critical component in directing the future? The Bank does suggest:

> Like trees, countries cannot be made to grow by being pulled upward from the outside; they must grow from within, from their own roots.
>
> (WB 1989: 194)

The Bank's arguments for local empowerment rest on two pillars: the first is an appeal to 'indigenous African values and institutions', particularly 'primary group loyalties', or a communal tradition (WB 1989: 60); the second concerns a debate which is couched in terms of not merely the public (state) sector *vis-à-vis* the private sector but also decentralization of institutional structure among central authority, local government and local communities (ibid.: 55) in order to ensure efficacy in development initiative.

It is worth quoting at length to illustrate the flavour of the Bank's proposals with respect to African 'tradition' and community initiative:

> History suggests that political legitimacy and consensus are a precondition for sustainable development. A sound strategy for development must take into account Africa's historical traditions and current realities. This implies above all a highly participatory approach – less top-down, more bottom-up than in the past – which effectively involves ordinary people, especially at the village level, in the decisions that directly affect their lives.
>
> (WB: 1989: 60)

'Africa has rich traditions of community and group welfare', the Bank writes:

> This is reflected in the widespread practice of sharing among people, with its emphasis on grassroots initiatives and community-based projects. Such cooperation tends to be spontaneous and informal.
>
> (WB 1989: 168)

African intellectuals, the Bank continues, have frequently noted the failure of post-independence development strategies to draw on 'the strengths of traditional societies' (WB 1989: 168). But, the Bank proposes: 'Communal culture, the participation of women in the economy, respect for nature – all these can be used in constructive ways' (ibid.).

The 'informal' credit system or indigenous methods of crop husbandry *inter alia* have a role to play. Without romanticizing 'the traditional sector', the Bank's purported aim is to support it through the 'modern' sector, rather than undermine it. Pragmatism is to prevail:

Community-based development projects provide an avenue for mobilizing community savings in cash or labour for a range of local activities. Much community development in Africa has been carried out by local self-help – for example, the construction, repair, and maintenance of community facilities. Because those involved are direct beneficiaries, motivation tends to be high. Such projects are an effective means of using free resources to meet the community's most urgent needs.

(WB 1989: 169)

Earlier in the report it is noted that:

[m]any basic services, including water supply, health care, and primary education, are best managed at the local level – even at the village level – with the central agencies providing only technical advice and specialized inputs.

(WB 1989: 54)

In order to facilitate local initiative, the aim should be, in the Bank's words: 'to empower ordinary people to take charge of their lives, to make communities more responsible for their development, and to make governments listen to their people' (ibid.). Later in the report, the following words are used: 'The challenge is to build on this solid indigenous base, with a bottom-up approach that places a premium on listening to people and on genuinely empowering the intended beneficiaries of any development program' (WB 1989: 191).

Empowerment is to be achieved, essentially, through 'a more pluralistic institutional structure' (WB 1989: 54), specifically through a devolution of authority from central to local government and through closer linkage with non-governmental organizations. The Bank argues as follows:

Involving central bureaucracies may stifle local initiatives. Thus NGO support structures should be encouraged to serve as a link between the state and thousands of small self-help and community development efforts. Such NGOs are closer than government to the rural communities, . . . highly motivated, cost conscious, and sympathetic to labor-intensive approaches, and flexible – a quality that stems from their small size and decentralized decision-making. Both local and foreign-based NGOs should be encouraged.

(WB 1989: 169)

The pervasive paternalism of the document aside, the assumptions in the above argument may be questioned on four fronts. First, the Bank draws on a notion of 'tradition' and communal solidarity, devoid of historical context, as an unchanging and unchallenged sphere of practice without ideological imputation in order to suggest prescriptions for future actions. But 'tradition' or 'custom', Glazier (1985: 231) suggests, is a 'continuously evolving code', used by individuals or groups of individuals to create strategies which 'serve economic or political ends' (ibid: 28). It is a malleable and manipulable instrument to which individuals from different genders, generations and classes have access (Mackenzie 1990; Stamp 1987) and which may be used to foster social differentiation (also Chanock 1985; Moore 1986; Parkin 1972).

This point is a critical one: societies may contain values and expertise far more consonant with a sustainable development ethic – including that of community responsibility – than that of a development model imported by outside agencies, but to assume that such societies are undifferentiated, that all interests are served through subscription to 'community development efforts' (WB 1989: 169), is to mythologize. Hyden (1980) notwithstanding, differentiation within the peasantry has created its own series of emergent and frequently conflictive interests (for example, Bernstein 1979; Kitching 1980), played out further in the reconstruction of gender (for example, Mackenzie 1990; Mikell 1989; Robertson and Berger 1986; Stamp 1987). The objectives of different interest groups may be served through community initiative, as Wade (1988) has illustrated with respect to collective decision-making *vis-à-vis* scarce resources in the very different socio-political context of 'village republics' in southern India, but this may not be universally assumed. Thomas (1988: 22) has argued, with reference to Harambee and the politics of

27

self-help in Kenya, that 'strong communal identities' frequently coexist with 'nascent class awareness'. Subscription to community solidarity may also coexist with contradictory gender awareness (Mackenzie 1990; Stamp 1987). The outcome may be the complex interplay of allegiances which in certain situations takes the form of collective community action while at other times the structure of such communities may be challenged from within.

Second, the Bank's prescriptions can be challenged for the assumption of 'free resources' (WB 1989: 169) that may be put to use to meet community needs. Labour no more than capital is a 'free resource' in the provision of basic needs: opportunity costs are implied in both. Again in Kenya, activities carried out under the aegis of Harambee do enable rural communities to command resources, through patrimonial linkage, from an urban élite with a need for a secure power base in the rural area, and thus at one level represent a form of progressive taxation. But they also serve to legitimize inequalities in the post Independence state (Thomas 1988), an argument which is pursued at the more general level by Brett (1988: 10). Thomas writes:

> The rhetoric of Harambee stresses cooperative effort for the benefit of all. The reality of Harambee underscores a system in which some communities, some groups, and some national-level elites benefit far more than others.
>
> (Thomas 1988: 23)

As Harambee activities frequently rely on women's labour (Thomas 1985), the stress on their contribution to community welfare may frequently be disproportionate to their gain. Continued reliance on their labour input in such activites becomes more problematical under structural adjustment programmes where increased responsibility for productive and reproductive tasks in rural areas falls to women, as discussed in the previous section.

This is not to argue that strategies created at the local level are not, and have not been, vital to survival in rural Africa, as several of the case studies illustrate. But it is to expose to scrutiny further curtailment of state responsibility.

A third area of concern for the Bank's new found panacea for Africa's ills lies in the assumption that 'local' empowerment is politically of no consequence to those in power at the level of the state. Here the Bank falls into the populism of early proposals for development 'from below' (see Forbes 1984: 130–5; Gore 1984: 222–32). Arguing for the ascendancy of territory rather than function as a key principle for social organization (for example, Friedmann and Weaver 1979; Friedmann 1981), such proposals focus on internal territorial integration through the

> maximum mobilisation of each area's natural, human, and institutional resources with the primary objective being the satisfaction of the basic needs of the inhabitants of that area.
>
> (Stöhr and Taylor 1981: 1)

In order to achieve equitable development, Stöhr and Taylor argue, 'development policies must be oriented directly towards the problems of poverty, and must be motivated and initially controlled from the bottom' (ibid.). Further,

> Development 'from below' strategies are basic-needs oriented, labour-intensive, small-scale, regional resource-based, often rural-centred, and argue for the use of 'appropriate' rather than 'highest' technology.
>
> (Stöhr and Taylor 1981: 1-2)

Selective regional, or spatial, closure (Stöhr and Tödtling 1979) is viewed as a critical instrument in achieving these ends.

But development 'from below' is labelled neo-populist by Gore (1984: 165) and utopian by Forbes (1984: 134) precisely because, although it may draw from political economy in considering, for example, the spatial transfer of surplus (Forbes 1984: 134) or the empowerment through participation of the poorest members of society, it fails to construct a coherent analysis which connects the different actors, or interest groups (class, gender, ethnic group), within society. As Forbes (1984: 134) points out, it is a theory of policy rather than of explanation. The contradictions that may evolve in terms of state–local relations, of classes, or of genders, are ignored. Indeed, the interests of individuals or of groups are frequently equated with territorial interest, 'an ecological fallacy' in Gore's (1984: 225) conception: it does not follow that development measured at a certain level of aggregation denotes benefits to all members of that group.

The empowerment of individuals or groups within a community concerns, by its nature, political action. The realization of strategic needs (Molyneux 1985), which go beyond those of 'coping mechanisms' or 'practical needs', reorders social relations. And these may be perceived as a threat, not only to local interests but also to state power. Nowhere is this clearer than in the response of the Kenyan state to the activities of the Kamiriithu Community Educational and Cultural Centre in Limuru. On two occasions, the first in 1977 after ten performances of the play *Ngaahika Ndeenda* (I will marry when I want), the second in 1982 after rehearsals for another play at the University of Nairobi, state authorities forcibly halted the performances, detaining writer Ngugi wa Thiong'o who had been involved in the production on the first occasion, and destroying the open air theatre in Limuru on the second.

Kidd (1982: 48) has asked why popular theatre, produced by workers and peasants, provoked the hostility of government: 'Why has a program which has significantly reduced illiteracy and alcoholism, increased employment opportunities, fostered a people's culture and raised the awareness and participation of villagers been suppressed?' The answer, in his subsequent analysis and in Ngugi's (n.d.) writings on the subject, identifies the theatre as a site of resistance, of struggle, and the dramas created as expressions of a growing conscientization of the people involved. 'Understandably', Ngugi (n.d: 126) writes, 'the wealthy who control the government did not like the stark realities of their own social origins enacted on the stage by simple villagers.'

Similar presumptions to those of the Bank are found in AAF–SAP, as has already been suggested. Where the state is a 'continual site of struggle' (Kitching 1985: 132), the primary concern of the contending interests which comprise the state is the reproduction of their own 'class' position (Dutkiewics and Shenton 1986). Thus, the political and economic restructuring necessary to facilitate empowerment may be unlikely to proceed far. With specific reference to the Berg Report and demands for a reduction in state intervention, Dutkiewics and Shenton (1986: 114) argue that the 'ruling groups' are in effect invited 'to commit political suicide'. Any strategy of decentralization in this context may be no more than a guise for the exertion of further central control (ibid: 111; Olowu 1989: 202. Gore 1984: 249–58). The 'metonymical' nature (Gore 1984: 258) of such policies allows the state both to resolve conflicting interests within it and to legitimate its continuation in authority.

To what extent, then, can local, non-governmental organization thrive? Can it provide, except under the most repressive regimes, resolution of problems of poverty, as Sandbrook (1986) claims? Is a threat to the rules of the game, which are operative in 'macro arenas' (Goulet 1989: 168), more or less likely to be tolerated as states retrench, in terms of the provision of social welfare, but increase repression in an attempt to enforce unpopular policies which are demanded through SAPs (namely the urban riots in Zambia in June 1990)? This fourth area of concern is neglected by both AAF–SAP and the World Bank.

The case for local empowerment may be strong – based on an ecological and environmental rationale (Richards 1979, 1983, 1985; Bell 1979; Chambers 1988; Thomas-Slayter and Ford 1989), on the reality of social and economic diversity (Chambers 1988; Taylor 1989; McCall 1988; Oakley and Marsden 1984), or on the ideology of a 'moral incentive' for equitable development and political democracy (Goulet 1989: 172). However, local organization, even if it originates in response to crisis, as 'the spontaneous mobilisation of a powerless group to defend itself against destruction', may yet contain, Goulet (1989: 167) argues, the 'latent seeds of organisation for multiple new developmental actions'. In this sense, the line drawn between that which may initially appear as a coping mechanism or action to meet a practical need and that which is of strategic significance may be blurred. And in the transfer of consciousness from one to the other, the transition to the 'macro arena' (the 'ever widening arena of decision making', which implies the maturation of a 'social force wielding a critical mass of participating communities' [Goulet 1989: 168]) will indeed challenge the interests of those in power both within the local community and at wider levels of social organization, the state.

NOTES

1 Gender is taken here to refer to the *social* construction of differences between women and men. Following Robertson and Berger (1986), the concept of access to and control of critical resources is used to analyse the relationships of men and women to the means of production. Such a definition of class, Robertson and Berger suggest (1986: 21), is 'more sensitive to local conditions and social structure' in Africa, and allows for an analysis of both class and gender struggle within and outside the household.

2 With respect to the Bank's classification of countries with 'strong' or 'weak' reform programmes (WB/UNDP 1989: 33, Note b), the UNECA (1989b) points to the lack of clearly developed criteria to distinguish among African countries. The Bank defines as 'strong' those countries with SAPs with the WB/IMF during 1985–7. But, the UNECA (1989b: 4) points out, 'neither the count of 19 countries resulting from this classification nor the reference period stated are adhered to in other parts of the Report'.

3 ECA calculations were based on World Bank GDP data (in constant 1980

US dollars, market prices) and countries were grouped according to those in the Bank's list. But the annual average growth rates for the groupings calculated by the ECA were based on weighted averages rather than the unweighted ones of the Bank (UNECA 1989b: 10).

4 These data differ from those given in AAF–SAP. Countries with strong SAPs are estimated in AAF–SAP to have had an overall average annual growth rate of 1.5 per cent in the period 1980–87. 'Weak-adjusting countries' recorded an overall average annual GDP growth rate of 1.2 per cent and 'non-adjusting countries' a rate of 3.1 per cent (UNECA 1989a: 22–3). These are the data used in Figure 1.1, and represent subsequent calculations to those in UNECA 1989b.

5 Abdoulaye Kone, Economy and Finance Minister, quoted in *Globe and Mail*, 2 July 1987 and referred to by Bienefeld 1989: 73.

2

LOCAL FARMER ORGANIZATIONS AND RURAL DEVELOPMENT IN ZIMBABWE

Lovemore M. Zinyama

INTRODUCTION

The political history of Zimbabwe from colonization in 1890 to the present has had a controlling influence on state policies for the development of those parts of the country that were set aside for settlement by the indigenous blacks *vis-à-vis* areas that were alienated for the incoming white settlers. Colonial state policies favoured the latter areas, thereby creating and perpetuating a dual socioeconomic structure which was characterized by wide spatial inequalities. The post-independence government that came to power in 1980 has sought to reduce these racial and spatial inequalities by giving priority to the development of the previously neglected areas that are home for the country's black peasant farmers who make up some 60 per cent of the national population. However, some commentators contend that recent gains in agricultural production and rural development, with their focus on the existing peasant areas but without substantial agrarian reform, are likely to be unsustainable. It is argued that the government's current policies and strategies have failed to address directly the problems facing the rural peasantry in Zimbabwe. These problems, which are a legacy of ninety years of colonial rule, include rural overpopulation and a deteriorating human–land ratio, environmental degradation, low rural incomes, and a lack of both rural and non-rural off-farm employment opportunities.

In the meantime, the small-scale subsistence agricultural population is compelled by harsh economic and environmental conditions to evolve a variety of strategies for mitigating the persistent rural poverty and hardships. This chapter examines some of the local-level initiatives that small-scale peasant farmers in Zimbabwe adopt in

33

order to improve their levels of agricultural production and hence their household incomes, with particular emphasis on the role of grass-roots voluntary farmer organizations. In many ways, these organizations complement the government's efforts to develop the previously neglected peasant farming areas. Given the unique circumstances of the decolonization process in Zimbabwe, where independence was won after a bitter war whose brunt was borne by the rural population, the development of these areas is seen as a two-way partnership in which the government contributes from the top while the farmers contribute from below. The focus of this chapter is on the latter, using the case of local peasant farmer groups as an example of 'development from within' or the 'self-reliant and communitarian' approach to rural development that has recently been suggested as one possible way of solving the African economic crisis (Mackenzie and Taylor 1989; Sandbrook 1986). Such community-based developmental organizations may

> also permit villagers, by defining their own needs, designing and implementing their own projects, to educate themselves in organizational dynamics and self-government. In time, this local capacity for organization and confidence may increase popular pressure for change at the territorial level.
>
> (Sandbrook 1986: 331)

The Zimbabwe experience in recent years provides an additional dimension to the theme of this book. In most African countries today, governments are retreating from the rural areas, leaving the rural population to evolve its own coping strategies and organizational structures in order to mitigate the deteriorating economic conditions. In Zimbabwe, on the other hand, the opposite prevails at present, with the government actively seeking to improve rural living standards through the development of social and economic infrastructure, the improvement of health and sanitation, and the provision of agricultural extension, marketing and credit services. Local farmer initiatives for the development of their areas and their agricultural capacities have generally received support from the government and, in some cases, from the private sector and from non-governmental organizations. Thus, in the Zimbabwean context, local-level initiatives or 'development from within' need to be regarded as complementary to the government's efforts of 'development from above'. It is therefore inappropriate to regard these local initiatives primarily as a reaction to the withdrawal of the state from the rural areas.

The chapter is divided into four parts. The first section briefly describes the principal agro-ecological regions into which Zimbabwe is divided for commercial farming purposes. The second section provides an overview and comparison of colonial and post-colonial government policies for the development of the black peasant farming areas. The third section discusses the four agricultural subsectors in the country and the changing role of black farmers in the national economy. The fourth section discusses how peasant farmers in two of the communal farming areas are using grass-roots voluntary organizations as a means of self-reliant resource mobilization in order to overcome felt agricultural resource constraints.

AGRO-ECOLOGICAL REGIONS OF ZIMBABWE

Zimbabwe has been divided for agricultural purposes into five agro-ecological regions (commonly known as Natural Farming Regions), mainly on the basis of the amount, reliability and variability of their rainfall (Vincent and Thomas 1961). Conditions become increasingly

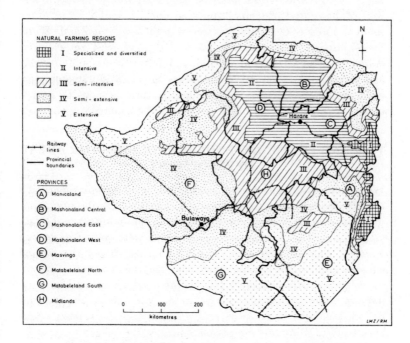

Figure 2.1 Principal agro-ecological regions of Zimbabwe

35

marginal for crop production from Natural Region I to Natural Region V as rainfall decreases in both amount and reliability. For commercial farming purposes, Regions I and II, with high and reliable annual rainfall in excess of 750 mm, are suitable for specialized farming (i.e. plantation crops such as tea and coffee) and intensive crop and livestock farming respectively (Figure 2.1). Agro-ecological Regions I and II respectively cover 1.8 per cent and 15 per cent of the country along the eastern border highlands and on the central plateau in the north and east. Region II is the breadbasket of the country, contributing up to 90 per cent of the national annual crop production.

Agro-ecological Region III receives moderate rainfall, between 650 and 800 mm per year, but much of it falls as infrequent heavy downpours which, combined with the high summer temperatures, reduce its effectiveness for crop production. The region is also subject to mid-season dry spells, thereby making it suitable for semi-intensive farming and marginal for enterprises based on dryland crop cultivation alone. Region III covers 18.7 per cent of the country, surrounding Natural Region II but with the largest expanse to the south and west of the central plateau. Natural Regions IV and V occur to the south and west of the country as well as in the low-lying Zambezi and Save-Limpopo valleys in the north and south east respectively. Annual rainfall is less than 650 mm with frequent seasonal droughts and severe mid-season dry spells. These areas are suitable only for extensive livestock and game ranching. Natural Regions IV and V occupy 37.8 per cent and 26.7 per cent of the country respectively.

PRE- AND POST-INDEPENDENCE POLICIES FOR RURAL DEVELOPMENT

When Zimbabwe attained independence in April 1980, it inherited a racially divided, dual socioeconomical structure. On the one hand, there was a modern sector comprising the urban-industrial and large-scale commercial farming subsectors which were controlled by the minority white settlers who made up less than 5 per cent of the national population. On the other hand, there was a predominantly subsistence agricultural subsector that provided a livelihood for some three-quarters of the black population. Throughout the colonial period, a variety of legislative and socioeconomic impediments had been used to minimize the permanent migration of blacks into the white-controlled areas, confining the men instead to oscillating labour migration and forcing them to sojourn in town while their families remained in the rural areas (Patel 1988; Zinyama and Whitlow 1986). At the 1982 census, only one-quarter of the black population was urban, a figure far below that of several other African countries at a comparable level of development. Even today, a decade after the colonial influx control measures were removed and after the country has experienced rapid urban growth arising from rural out-migration, the rural areas continue to support the majority of the population. It is to these rural people that the post-independence government is giving priority in the allocation of national development resources.

Following colonization of the country in 1890 by the British South Africa Company (BSACo), land was divided along racial lines in such a way that blacks were mostly confined to agriculturally marginal areas in Natural Regions IV and V. In 1969, with the enactment of the Land Tenure Act, the last major piece of colonial legislation for dividing the land, the country was divided equally between blacks and whites, with each group getting 46.6 per cent of the total land area, and the remainder was designated as national parks and forest areas (Christopher 1971; Floyd 1962). The areas that were set aside for blacks (previously called Native or African Reserves and now Communal Lands) not only experience low and unreliable rainfall but also have poor sandy soils derived from granitic rock formations as well as large tracts occupied by exposed granite domes which reduce the amount of land available for cultivation (Whitlow 1980a, 1980b). Three-quarters of the communal farming areas are situated within Natural Regions IV and V, compared with 49 per cent of the former white large-scale commercial farming areas. Almost two-thirds of the land in Natural Region I and three-quarters of that in Region II falls within the previously white-owned large-scale commercial farming sector.

In the early years after colonization, blacks were able, by increasing their agricultural output, to meet not only their new tax liabilities, imposed by the colonial administration, and other emerging requirements for cash but also the growing demand for food in the newly established towns and mining settlements. However, as the settler agricultural sector became more firmly established with active government support during the first two decades of this century, it became necessary, first, to ensure a source of cheap labour for white farms and mines and, second, to protect the agricultural produce markets, for the benefit of white farmers, by eliminating competition from black farmers. The government instituted a series of measures which restricted access by black farmers to agricultural input and output markets and support services (i.e. research, extension and credit facilities). This combined with diminishing productivity from impoverished soils in increasingly overcrowded reserves, reduced black farmers' net returns from agriculture and compelled them instead to sell their labour on the white-owned farms, in the mines and in the towns. The communal areas henceforth served as cheap labour reserves for the modern white-controlled sector of the national economy. Meanwhile, increasing overpopulation within the communal areas over the years has exacerbated problems of land shortage, of the cultivation of marginal lands on steep hill slopes and long stream banks, and of general land degradation through soil erosion and deforestation (Whitlow 1980a, 1988). Not surprisingly, therefore, black farmers during the 1970s contributed less than 25 per cent of the national annual agricultural output while over 90 per cent by value of all of the agricultural produce marketed through official marketing boards came from some 5,000 white large-scale commercial farmers (Republic of Zimbabwe 1982).

At independence in 1980, the new government came to power committed to transforming the national economy and society to socialism and, as a more immediate task, to redress the inherited social and spatial inequalities. It immediately embarked on a dual rural development strategy involving, on the one hand, the redistribution of land and resettlement of land-hungry peasants on former white-owned large-scale farms and, on the other, increased state support for farmers within the existing communal areas.

The government had originally planned to resettle some 162,000 families by 1985. However, by the middle of 1989, only some 52,000 families had been resettled on nearly 3 million hectares of land under different settlement and land use models (*The Herald*, 4 August 1989; Kinsey 1983; Zinyama 1986a). The resettlement programme has progressed much slower than originally envisaged for a number of reasons, including: constitutional constraints on land acquisition, which were imposed on the government at the Lancaster House conference in London prior to independence; lack of money for land purchase and development; recurrent droughts during the mid-1980s; the continuing dominance of settler economic planning ideology; and the co-optation of newly affluent and influential blacks into the land-owning classes (Munslow 1985; Stoneman and Cliffe 1989; Weiner 1988; Zinyama *et al.* 1990). Although the resettlement programme was aimed at redressing inequalities in the distribution of land, it has effected little change in terms of both the total amount and the agricultural potential of the land that has been transferred to date. Much of the land that has so far been transferred for resettlement is situated in agro-ecological Regions III, IV and V. Furthermore, the programme has been scaled down considerably. The 1986–90 National Development Plan proposed the resettlement of only 15,000 families per year (Republic of Zimbabwe 1986). Even this reduced target of 75,000 settler families over five years is unlikely to be achieved. (The constitutional restrictions on land acquisition fell away in 1990 and it still remains to be seen whether the pace of the resettlement programme will accelerate.)

The government has been more successful in the measures that have been taken to improve agriculture and general living conditions within the communal areas. These measures include: the provision of schools, health services, and improved water and sanitation facilities; the decentralization of government services; and improved accessibility of such services through the development of rural service centres at district level (Gasper 1988; Wekwete 1988; Zinyama 1987a). Measures taken to improve the farmers' levels of agricultural production, and hence on-farm incomes from sales of crop surpluses, include: the establishment of a network of official marketing depots covering most of the rural areas; the provision of agricultural credit and extension services; and the improvement of the rural road network (Zinyama 1986a). Local-level participation in the identification, articulation and implementation of development projects within the communal areas has been facilitated by the democratization of local government and the establishment of a hierarchy of development planning fora at which the local inhabitants are represented in Village, Ward and District Development Committees (VIDCO, WADCO and DDC), up to the Provincial Development Committees (PDC) (Mutizwa-Mangiza 1986)

THE CHANGING ROLE OF BLACK PEASANT FARMERS

Today, four distinct agricultural subsectors exist in Zimbabwe. These are: (a) the large-scale commercial farming sector; (b) the small-scale commercial farming sector; (c) the communal farming sector; (d) the resettlement areas (Figure 2.2).

The large-scale commercial farming sector occupies some 12.8 million hectares of land, covering approximately 33 per cent of the total land area. The sector comprises some 4,500 farming units which

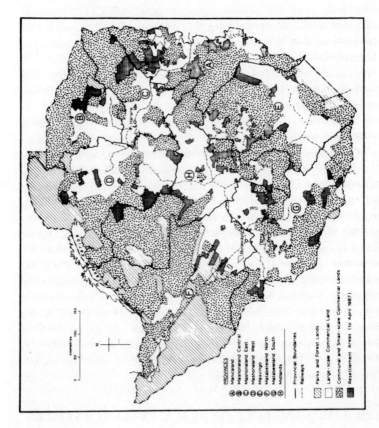

Figure 2.2 Zimbabwe: major land use divisions

range in size from less than 200 to over 8,000 hectares, mostly under freehold tenure. These farms were previously exclusively owned by whites, but today some wealthy blacks have acquired properties within this sector.

The small-scale commercial farming sector comprises some 8,500 units on land that was set aside during the colonial period for the few blacks who wished and could afford to purchase land under freehold and leasehold tenure. The farms are generally less than 150 hectares in size. The sector occupies some 1.4 million hectares, or 3.6 per cent of the total land area.

The communal farming sector occupies some 16.4 million hectares of land, or 42 per cent of the total land area. The sector supports, directly from agriculture, about 60 per cent of the national population. The average size of arable holdings in the communal areas is about two hectares, held under usufructuary rights. In addition, farmers have access to communal grazing land.

The fourth subsector, the resettlement areas, is a post-independence development. The resettlement areas occupy about 8 per cent of the national land area at present. As mentioned earlier, much of the resettlement land has been at the expense of the large-scale commercial farming sector, although some vacant state land has also been utilized for the resettlement programme.

Since 1980, black small-scale farmers in the communal, small-scale commercial, and resettlement areas have greatly increased their crop production levels as well as their share of total marketed agricultural produce as a result of the support and incentives given by the government. This is particularly true for maize and cotton, their principal cash crops. Maize is also the staple crop in the country. Until 1980, black farmers accounted for less than 10 per cent of the maize and less than 25 per cent of the seed cotton that was sold annually to the respective official marketing organizations, the Grain Marketing Board (GMB) and the Cotton Marketing Board (CMB). Today the black small-scale farmers account for a little over 50 per cent of both the maize and cotton that is sold annually to the respective marketing boards. Sales of maize to the GMB from black farmers have increased sixfold from a pre-independence peak of 84,300 tonnes, which were sold after the 1977–8 season, to over half a million tonnes per season today. These figures exclude retentions for the farmers' subsistence requirements. Similarly, the amount of seed cotton that is sold to the CMB has more than doubled from less than 45,000 tonnes annually before independence to over 100,000 tonnes per year today

41

(Agricultural Marketing Authority 1988; Cotton Marketing Board 1988; Grain Marketing Board 1988).

Such progress can only be described as remarkable, especially when viewed against a backdrop of the recurrent severe droughts of the mid-1980s. Many observers have commented on this apparent Zimbabwean success story in transforming the peasant agricultural sector within such a short period of time. However, these figures, while certainly impressive at the national scale, hide substantial spatial and social disparities in the transformation of the country's peasant agricultural sector. The continuing maldistribution of land across the agro-ecological regions, which was described earlier, exerts a considerable differentiating influence on the current spatial development pattern of the peasant sector (Stanning 1987; Zinyama 1988a). For instance, the three provinces of Mashonaland Central, East and West, which are located in agriculturally favourable areas on the central plateau, support about 28 per cent of the total black farming population. The communal, small-scale commercial, and resettlement areas within these three provinces have experienced the highest and most consistent increases in crop sales to the marketing boards since 1980 (Zinyama 1988a). In seasons of normal rainfall, the three provinces contribute between 55 and 65 per cent of the total maize that is sold by black small-scale farmers countrywide. This rises to over 95 per cent following severe droughts when lower rainfall areas are hardest hit. Some two-thirds of all of the maize that was sold to the GMB by black small-scale farmers during the period 1980–86 came from a little over one-quarter of the farming population who are fortunate to live in these three provinces. An even greater pattern of inequality emerges with respect to sales of seed cotton to the CMB. Some two-thirds of the annual cotton output of the small-scale farming sector (i.e. equivalent to 20–25 per cent of the total national output) comes from two major producing areas in the north west, in the Midlands province, and in the north, in the Mashonaland Central and West provinces. The comparative study by Rohrbach (1987) of two communal areas, Mangwende, which is 80 km east of Harare in Natural Region II, and Chivi, some 370 km to the south in Natural Region IV, also demonstrated that the relative impact of any policy or institutional intervention depends very much on the agro-ecological conditions of the areas concerned.

Intra-community differences by age, gender and status have also been reported in other studies of the peasant agricultural sector. For instance, in Wedza communal area south of Harare, Callear (1984) found three categories of households which were particularly disadvantaged: (a) young families with little land, mainly because of growing population pressures, with few or no cattle and with few or no implements because they had not yet had time to accumulate the necessary farming assets; (b) widows, especially if they had school-going children whose daily labour time was eroded by schooling; (c) families of some labour migrants. The three categories of families failed to earn adequate incomes from both on-farm and off-farm activities. Similar intra-community disparities were reported by Gobbins and Prankard (1983) from four communal areas north-west of Harare, in the Mashonaland West province. In their study, Gobbins and Prankard (1983) classified farming households on the basis of whether or not a family member had undergone formal on-farm training in agriculture under a government extension worker, thereafter qualifying to be known as a 'master' farmer. In all four areas, master farmers had access to more land and family labour, owned more draught cattle, used more purchased crop inputs, particularly fertilizers, and generally had a more balanced mix of resources than non-master farmers. The former also obtained higher yields and incomes from crop sales. The authors suggested that it was members of families who were already wealthy who trained to become master farmers, thereby adding formal technical knowledge to their existing favourable resource base. The result would be an accentuation of intra-community socioeconomic inequalities.

Overall, the regional and social disparities emerging from the commercialization of peasant agriculture suggest that on-farm incomes for a large proportion of communal farmers, especially in the low rainfall areas, have changed little from what they were before 1980. The benefits from the new rural development thrust are largely accruing to the small number of peasant farmers who are fortunate enough to be located within the better agro-ecological regions. Elsewhere, farmers continue to be handicapped by the constant threat of drought and food shortages, by increasing land shortages, by infertile soils and by low agricultural production. Given the social and regional disparities in food production and distribution, it is not surprising that malnutrition continues to be a major public health problem in many parts of rural Zimbabwe (Republic of Zimbabwe 1986), in particular in low rainfall areas where a considerable proportion of children are undernourished even during supposedly good crop years. Overall, post-independence agrarian restructuring has been limited. The success to date in raising agricultural productivity

43

within the communal areas appears to be merely short term; it does not provide a long term solution to the problems of rural development in Zimbabwe (Munslow 1985; Weiner 1988; Zinyama *et al.* 1990).

Government efforts to develop the rural areas of Zimbabwe, as discussed above, represent the 'development from above' approach. The success in implementing these programmes has been made possible partly by the unique circumstances surrounding the decolonization of the country, whereby the liberation war in the 1970s was fiercest in the rural areas, so enabling the nationalist forces to mobilize and identify with the rural population. Thus, the priority given to rural development after independence was the new government's way of thanking the rural populace for its support during the war. The remainder of this chapter will focus on some of the ways in which farmers themselves are contributing towards the improvement of both agricultural production and on-farm incomes through sales of surplus crops in two of the three communal areas: a process of 'development from within' which, as mentioned earlier, is complementary rather than being an alternative to the macro-level strategies and programmes of government.

LOCAL ORGANIZATIONS AND RURAL DEVELOPMENT: A CASE STUDY

The study areas

During 1983–4, a survey was undertaken by the author to examine the factors influencing agricultural change and development in two communal areas, namely Mhondoro and Save North (Zinyama 1988c). The former is situated about 40 km south-west of Harare while the latter is 150 km south-west of the capital (Figure 2.3). Mhondoro is therefore more accessible than Save North to national markets, infrastructure and other agricultural support services which are available within the capital city and other urban centres. At the 1982 census, Mhondoro had a population of 59,800 people in some 10,900 households while Save North had 62,900 people in 12,600 households. In Mhondoro, population densities ranged between 23 and 107 persons per km^2 for individual enumeration areas, with an average density of 46 per km^2. Average density in Save North in 1982 was 48 per km^2, varying from less than 20 to a little over 100 per km^2 for individual enumeration areas. In Mhondoro, 430 randomly selected households (4 per cent sample) were interviewed. In Save North, 371 households (3 per cent sample) were interviewed.

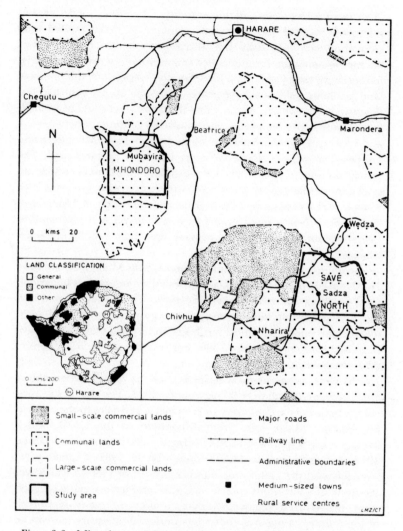

Figure 2.3 Mhondoro and Save North communal areas: the survey areas
and major land use divisions south of Harare

The two communal areas are situated in Natural Farming Region
III, with moderate rainfall of 650–800 mm per year. Soils in both
areas are largely sandy, derived mainly from granitic rock formations. Although the soils are light and relatively easy to cultivate, they
are inherently infertile, they drain freely and dry out quickly. This

45

gives them a low water holding capacity. Therefore, moderate and infrequent rainfall and poor soils combine to make rainfed crop cultivation not only risky but also expensive where the farmers attempt to apply the recommended quantities of inorganic fertilizers for optimum crop yields. In both areas, the average amount of arable land per household was 2.1 hectares, although 60 per cent had less than two hectares each. Another one-third had what could be termed, by communal area standards, medium-sized holdings of between 2.1 and 4 hectares. Only 5.4 per cent of the households in Mhondoro and 1.9 per cent in Save North had more than four hectares each. The principal crops in both areas are maize and groundnuts, which are grown by virtually every household, with maize being grown as both a staple and a cash crop. Minor crops include cotton in Mhondoro, and in Save North small grains such as bulrush millet (*Pennisetum typhoides*), sorghum (*Sorghum vulgare*) and finger millet (*Eleusine coracana*).

In common with other communal areas, both Mhondoro and Save North have, since independence, witnessed major changes in agricultural practices and crop production levels. The survey showed that the proportions of households using improved agricultural techniques, particularly chemical fertilizers and hybrid maize seed, have increased considerably since 1980. For instance, only 33 per cent of the households in Mhondoro (and an even lower percentage in Save North) were using chemical fertilizers for maize production before 1980, compared with over 90 per cent by 1983. The changes in agricultural practices are reflected in crop yields and sales to the marketing boards. Maize sales from Mhondoro to the GMB, which averaged 1,014 tonnes annually during the pre-independence period 1974–9, quickly rose to 14,038 tonnes for the 1981–2 intake year. This followed the nationwide bumper harvest of 1980–1 which was the result of a fortuitous coincidence of several factors, namely: good rains; the return to relative peace and improved security over much of the country; a massive government pre-planting price increase of 41 per cent to Z$ 120 per tonne for top grade maize; and the distribution of free seed and fertilizer packs as part of an internationally supported programme of post-war rural reconstruction and rehabilitation. There was a drop in sales to pre-1980 levels during the next two intake years, 1982–4, because of severe droughts during the 1981–3 seasons. However, maize deliveries from the area reached a new peak of 15,112 tonnes in 1984–5, before dropping to 12,174 tonnes for the 1986–7 intake because of poor and erratic rainfall during the previous 1985–6 growing season. Similar trends, though with lower figures, have occurred in Save North where maize sales to the GMB have increased from a pre-independence annual average of 95 tonnes during the period 1974–9 to 7,868 tonnes in 1984–5 and 5,601 tonnes for the 1986–7 intake year.

However, these aggregate figures hide substantial age, status and gender differences in crop production and sales. Generally, households with permanently resident middle-aged male heads performed better than households with either female heads or those with younger, absent, male heads. Households in the former category had more land, more farm equipment and more draught cattle than the other two categories. A large proportion of the crops that were sold in each area came from a small percentage of the sample farming population. For instance, in Mhondoro, 25 per cent of the households were not selling any maize while another 22 per cent sold less than ten bags (one bag equals 91 kg). Only a little over 10 per cent of the households sold over 50 bags of maize annually. In Save North, the proportion of households who sold little or no maize was even greater, as much as 75 per cent in a fairly good season, while less than 3 per cent sold more than 50 bags.

The respondents were asked what they considered to be the major obstacles that were hindering them from increasing their crop production. In both areas, the five most frequently cited constraints related to the lack of adequate farming resources. These were, in descending order of magnitude: (a) lack of money to purchase farm inputs, particularly fertilizers; (b) lack of farming implements, especially the ox-drawn single furrow plough; (c) lack of draught cattle; (d) inadequate family labour (Zinyama 1988b). An understanding of these constraints, as perceived by the farmers themselves, is essential if one is to understand the nature of local group activities in which farmers participate voluntarily, thus giving such organizations greater prospects for success.

Local farmer organizations in rural development

One of the strategies that is being used by the farmers to raise their level of crop production and on-farm incomes is participation in local voluntary organizations for collective action, as the farmers seek to overcome their felt resource constraints. Given the relative weaknesses of peasant households when they act individually for purposes of resource mobilization and acquisition or when they deal with the state or other external agencies, collective local action becomes a vital strategy for rural development. Moreover, the process of rural development entails increasing the participation of the people concerned in the decision-making process, and this can be enhanced through local groups. Local farmer groups can also be used to facilitate the collective purchasing and transport of agricultural inputs, marketing of produce, and the timely mobilization of labour for a variety of tasks such as ploughing, planting or weeding.

The distinguishing feature of these collective local action groups is self-management. Where the group is formally structured, management will be done through a committee which is elected by the farmers from among themselves. Self-management sets these groups apart from externally sponsored organizations which rely on governmental or non-governmental professionals and technocrats not only for advice but also for leadership. The principles governing such farmer groups are therefore voluntary membership, government by agreement, and social control through peer pressure (Bratton 1986, 1987). The problems that can beset group projects which are initiated from above without prior involvement of the target population have been clearly documented by Sibanda (1986) in a study of a goat breeding project that was intended to benefit the women of Muzarabani Communal Land in the Zambezi Valley in northern Zimbabwe.

Bratton further argued, as a result of his work elsewhere in Zimbabwe, that farmer groups are most likely to form and to succeed in those parts of the country where services from the various economic sectors, both state and private, are well represented and that they are least likely where the latter are weakly represented. He observed that:

> Autonomous farmer organizations are most likely to arise where effective state and market institutions provide stimulation and support. It is also vital that the policy regime allows sufficient 'space' for unaffiliated and autonomous organizations to exist.
>
> (Bratton 1987: 226)

The importance of state support for local voluntary organizations, particularly during their formative stages, was also highlighted by Brett (1988). He noted that the development of autonomous local organizations was not a substitute for effective public service provision, 'but only possible where it can grow effectively in association with it' (Brett 1988: 10). It can therefore be hypothesized that, given the current favourable national policy environment for local-level co-operative action, such organizations will be better developed and more effective in Mhondoro than in Save North because of the former's greater accessibility to the national economic heartland and larger marketable crop surpluses.

Mutual-help groups

A number of local voluntary farmer groups, traditional and modern, informal and formal, are found in both Mhondoro and Save North. First, there are the mutual-help groups which provide a means for overcoming labour and other resource constraints that households may face. The informal type of mutual-help group (called a *nhimbe* in the vernacular) is where a farmer provides beer and food in return for the collective labour of friends and neighbours on a particular day and for a specific task. The more formal type (*jangano*) is where a small number of friends or relatives agree to help each other by combining their labour on a regular rotational basis for such tasks as ploughing, planting, weeding or harvesting. In the former, both males and females pool their labour and other resources for the day. In the latter case, men usually work together mainly at ploughing only; the more prolonged activities such as weeding or harvesting usually involve women and children working co-operatively.

Analysis of the demographic structure of households in the two areas showed that those with female heads (i.e. widows and divorcees) accounted for 21 per cent and 14 per cent of all households in Mhondoro and Save North respectively. Although the remainder were headed by males, two-fifths of the male heads of households lived away from home as migrant workers. Thus, the agricultural labour input, day-to-day management and routine decision-making on the family farm were the responsibility of women in a large majority of households. Household labour was further reduced by the fact that two-fifths of the *de facto* or resident population were children attending local schools. Therefore, although the average household size for the resident population alone was 5.3 in Mhondoro and 5.5 in Save North, the actual amount of full-time labour which was usually available for agriculture was only two persons per household in both areas (Zinyama 1986b). Again, households with female heads and those where the male head was absent (i.e. female-managed households) generally had less family labour than those with resident male heads. Given the high incidence of male absenteeism and the post-independence expansion of educational opportunities for rural children, it is therefore not surprising that farmers identified the shortage of family labour as one of the principal constraints against increasing their crop production. One way in which they were trying to overcome this problem was through the formation of mutual-help groups.

Mutual-help groups involving the pooling of scarce production resources also help to improve access to essential production assets and implements among rural households, particularly where inequalities exist in the ownership of resources. For instance, some 30 per cent of the households in both Mhondoro and Save North did not own an ox-drawn single furrow plough, the standard implement for land preparation in communal area agriculture. A little over one-third had no draught cattle, and a further one-fifth had between one and two animals only to provide draught power. The latter group of households were in almost as precarious a position at the end of the long dry season, which coincides with the peak demand period for draught power for land preparation (November–December). Overall, a little over half of the households in both areas could be said to suffer, to varying degrees, from a shortage of draught power. Only 56.7 per cent of households in Mhondoro and 59.6 per cent in Save North had both ploughs and draught cattle of their own. One-quarter of the households in each area shared either a plough or oxen, usually with other members of the extended family, for which no payment was made. Another 18.2 per cent in Mhondoro and 10.8 per cent in Save North hired ploughs and/or draught cattle from other families, with payment being made in cash or in kind. Although the practice of sharing helps to improve access to resources, it also means that land preparation cannot be completed on time for all households, with a consequent loss of yield because of failure to take advantage of the early season rains. The most disadvantaged households with respect to ownership of farming implements and livestock were, first, female-headed households and, second, newly established households with younger male heads who have not yet accumulated an adequate range of assets (Zinyama 1987b).

Farmer training groups

It is now the general policy of government agricultural extension staff to work with groups of farmers rather than visit them individually on their landholdings. Given the current poor ratio of farmers to extension staff, the group approach enables extension workers to reach more farmers than would otherwise be the case through individual farm visits. Interested farmers within an area organize themselves into groups that meet regularly with the local extension worker for training and information dissemination. These groups are called Master Farmers' Clubs. A 'master farmer' is one who has successfully completed a prescribed programme of on-farm training under the supervision of the agricultural extension staff. In mid-1983, there were 96 master farmers' clubs with 4,671 members in Mhondoro, an average of 49 members per group, both males and females. The primary function of these groups is to train farmers and to improve the farmers' standards of agriculture. It is important to emphasize that, although the farmers receive technical support from government extension staff, the formation of the groups is largely initiated by the farmers themselves who are also responsible for their management through committees which are elected from among themselves.

Membership of master farmers' clubs is open to all interested farmers, including prospective trainees, regardless of gender, age or status. In both Mhondoro and Save North, membership was divided more or less equally between males and females. Male members are usually drawn from those who are permanently resident at home. Many of the female members are wives of absent migrant workers (female farm managers). Equality of membership in such skills training groups is being actively encouraged by the government as part of its efforts to raise the social and economic status of women in both rural and urban areas. Thus, a similar pattern was reported by Smith (1987) in her study of the participation of women in producer co-operatives in Zimbabwe. Although the research in Mhondoro and Save North did not examine the participation of women in decision-making within these groups, the observations by Smith for the producer co-operatives seem to be applicable to the communal areas as well. She noted that women, although enjoying equality in terms of membership, participated less than men in decision-making either as members of management committees or through contributions at general meetings. Reasons why women are under-represented in decision-making include: heavy domestic workloads and other socio-cultural demands which prevent them from serving in elected positions and which reduce their ability to travel and to represent their organizations externally, lack of administrative experience;

51

attitudes and prejudices against the election of women to positions of leadership.

Membership of farmer training groups is also differentiated in terms of social status and age. First, members tend to be drawn from the middle and upper social strata of the rural population, that is from households who already possess adequate land, farming implements and cattle. They are also more likely to have greater access to sources of cash (e.g. from crop sales, from loans from the Agricultural Finance Corporation, or from migrant labour remittances) with which to purchase farm inputs such as hybrid seed and chemical fertilizers. These groups therefore exclude the poorest of the farming population. Second, the majority of members are drawn from middle-aged and older households rather than from younger, more recently established families. The latter have fewer agricultural resources and, where the husband is employed elsewhere, the wives tend to have a more negative attitude towards farming (Zinyama 1988b). These wives frequently spend most of the dry season away from the family farm with their husbands, returning only at the onset of the rains. They are therefore unable to participate in much of the dry season training activities, in land preparation or in collective input procurement with other farmers.

Agricultural marketing groups

The third category of local farmer groups are those which are engaged in the purchasing of agricultural inputs and in the marketing of commodities to the statutory marketing boards. Membership of these marketing groups frequently overlaps with that of master farmers' clubs. This category includes voluntary savings or thrift clubs which are initiated and run by the farmers themselves with the assistance of their agricultural extension workers and, in some cases, with that of the country's two fertilizer companies. In Mhondoro, other savings groups are supported by the charitable organization, the Catholic Association, under its Silveira House Agricultural Project, as part of its programme of 'mushandira pamwe' (working together) groups, and by the multinational agricultural chemical company, Ciba-Geigy, under its 'kohwa pakuru' (expand your harvest) programme. The Catholic Association and Ciba-Geigy have been operating in Mhondoro since the early 1970s and 1981 respectively. Both distribute input packages either on loan (now using funds obtained from the Agricultural Finance Corporation) or for cash and they require that farmers work in groups on each task such as planting, weeding, harvesting and marketing. This obviously reduces the unit costs of loan administration, particularly where it involves the processing of large numbers of small loan requirements from the farmers.

Members of savings clubs hold regular meetings at which they deposit money into a club fund. The money is then used for purchasing inputs collectively for the next agricultural season. The quantities of inputs that are ordered by each member will depend on his/her requirements and on the amount of money that has been deposited with the club. At least two major cost advantages accrue to farmers who are members of these clubs. First, by obtaining their fertilizers directly from manufacturers in bulk at wholesale prices, which also carry cash discounts, the farmers benefit from lower purchasing costs. Second, the companies will then deliver the fertilizer to an agreed collection point (e.g. the home of the club chairperson) at much lower cost than if the farmers transported their inputs privately by hired lorry or on buses. During the 1982–3 and 1983–4 seasons, farmers who were members of savings clubs in Mhondoro paid between 15 and 20 per cent less for each 50 kg bag of chemical fertilizers, inclusive of transport, than non-members who purchased their requirements individually either at their local business centres or from the towns. (The fertilizer companies can also arrange to obtain and deliver hybrid maize seed and other chemicals for the farmers at the same time as they deliver the fertilizers.)

Farmers' groups for purchasing inputs are more numerous and have operated for longer in Mhondoro than in Save North. Some of the extension workers in Save North made initial attempts during 1982 to organize and convert some of the master farmers' clubs into savings clubs for input purchasing. A few of the farmers purchased their inputs through the clubs for the 1982–3 season. However, output that year was so severely cut because of the drought that most farmers were unable to obtain money to continue their club subscriptions in preparation for the next season (1983–4). Thus, in 1984, only 7 per cent of the households interviewed in Save North were members of savings clubs (most of which were largely inactive), another 4 per cent had ceased to be members, and the remainder were not members at all. In more recent years, the number of savings clubs has increased, mainly through the addition of functions to existing master farmers' clubs. In Mhondoro, a few of the savings clubs were started in the early 1970s but most of those operating today were formed after 1980. By mid-1983, at the time of the survey, there were 88 clubs with 4,313 members throughout Mhondoro, an average of 49 members per club.

A brief outline of two of the most successful savings clubs in Mhondoro is indicative. Savings club 'A' was formed in May 1982 with 35 members. By October of that year, membership had risen to 63. The members had purchased chemical fertilizers valued at $Z 3,843, an average of Z$ 61 per member, in preparation for the 1982–3 season. Savings club 'B', the oldest in Mhondoro, was formed in 1972 with 37 members. In September 1983 it had 107 members. By September 1983, the members had already bought Z$ 7,970 worth of fertilizers for the 1983–4 season. The club had also just completed, out of its own resources, the construction of a large brick-under-asbestos building which was to be used as a club office and a farmer training centre, as well as providing storage space for agricultural materials.

The varying history of collective groups in Mhondoro and Save North, as outlined above, accounts for variations in the impact of these groups as agents of change in the two areas. In Mhondoro, 47.2 per cent of the households were members of savings clubs, through which they purchased their inputs, compared with only 6.7 per cent in Save North. These savings clubs provide a vital vehicle by means of which farmers in the communal areas, given their limited individual input requirements and lack of money, can reduce costs of purchased inputs by buying collectively in bulk. Membership of such clubs also suggests that the farmers are looking beyond the traditional inputs of unimproved maize seed and cattle manure and that they are searching for alternative modern inputs which will give them greater yields, more marketable surpluses and higher incomes from farming.

It was therefore not surprising to find that farmers who were members of clubs not only applied more appropriate quantities of chemical fertilizers to their crops but also produced and sold more than non-members. The importance of appropriate fertilizer applications was highlighted by Tattersfield (1982). Using data from the large-scale farming sector, he estimated that proper application of the full package of modern technology, derived from research, would have resulted in an increase in maize output per hectare of 325 per cent over the output for 1950, with almost two-thirds of that increase coming from the application of nitrogenous fertilizers and a little under one-fifth from the use of hybrid maize seed varieties. The survey in Mhondoro showed that three-quarters of the households who used little (under 50 kg) or no chemical fertilizers on their maize crop during the 1981–2 season harvested less than ten bags of maize each (one bag equals 91 kg), and only 4.5 per cent of them gathered more than 30 bags. On the other hand, of the 121 households who used more than 300 kg of chemical fertilizers, 82 per cent obtained more than 30 bags each, with 13 households harvesting over 100 bags each. (The recommended minimum levels for maize are 250–300 kg per hectare, i.e. 5–6 bags of 50 kg each, comprising three bags of basal and two bags of top dressing nitrogenous compound. The average area that was planted with maize was 1.1 hectares.) About 50 per cent of the households that were members of savings clubs used more than 300 kg of fertilizers for their maize production, compared with only 14 per cent of the households that were not members. Out of the 76 households that harvested more than 50 bags of maize in Mhondoro, 77 per cent were members of savings clubs.

Similar results were reported by Bratton (1986) from four other communal areas of Chipuriro and Dande in the north of the country, Wedza in the centre and Gutu in the south. He found that farmers working in groups consistently produced more, both per unit of land and per household, and sold more than farmers working alone. More significantly, he found that farmers who were organized into groups for purposes of input acquisition performed better than those who merely pooled their resources of labour or draught power at the village level. The former performed better because they were more likely than the latter to receive the full package of agricultural support services, namely extension advice, fertilizers, hybrid seed, and market access.

Overall, during the past decade, farmers in Mhondoro have been more successful in transforming their agricultural practices and in raising average productivity levels than farmers in Save North. While these emerging regional differences are attributable to a number of factors, the existence of viable savings or thrift clubs, which farmers can use to mobilize effectively local capital resources, has certainly been a major contributing factor towards the greater agricultual success in Mhondoro.

CONCLUSION: SOME POLICY CONSIDERATIONS

This chapter started by examining the current macroeconomic environment for rural development in Zimbabwe. Unlike other African countries where national policies are generally unfavourable towards the rural sector, thereby adding to its stagnation, the rural areas in Zimbabwe are fortunate in that the government is currently giving them high priority. There is therefore considerable complementarity between the efforts of the state and those of the rural population itself to develop the rural sector. To illustrate the latter's contribution, the second part of the chapter focused on the role of local voluntary farmer organizations and on how they can provide an effective grass-roots vehicle for overcoming constraints that are identified by the farmers themselves as hindering their agricultural activities. In Mhondoro and, to a lesser extent, Save North, these farmer groups are helping to overcome problems arising from household labour shortages and from inequalities in access or ownership of agricultural resources. Such group activities should be actively encouraged; at the same time, they should not be strangled by restrictive rules and regulations nor have their principal advantages of voluntary membership, self-government and peer control removed from them.

Lack of money with which to purchase agricultural inputs, particularly fertilizers, was the most frequently cited constraint faced by farmers in both Mhondoro and Save North (Zinyama 1988b). While the provision of institutional credit for small-scale farmers by the Agricultural Finance Corporation (AFC) since 1978 has gone some way to help them, a number of problems remain. For example, 90 per cent of the loans are short term or seasonal for crop production purposes; the AFC tends to favour farmers in the wetter regions who are able to present viable cropping programmes when they submit their applications for loans (Zinyama 1988a). The frequent droughts that are experienced in low rainfall regions, where the majority of communal area farmers live, reduce the ability of farmers to benefit from the loans because they are then unable to repay them. Some farmers who have accumulated substantial debts are now evading repayment by selling their crops indirectly through GMB-approved grain buyers or through other farmers. As a result, in 1988 the AFC announced a change in policy: to move gradually from individual to group lending in order to reduce the problem of non-payment by small-scale farmers. However, the proposed debt collection system involving mandatory joint liability can also give rise to other problems. For instance, the whole group may opt not to repay its loans to the AFC: or, because the lending agency will automatically deduct the debts of defaulting members from the accounts of paying members, a situation of 'internal indebtedness' may arise within groups. Whether credit is given under individual liability or mandatory group liability, the risk of increasing rural indebtedness is very real. The recurrent droughts in low rainfall regions and the likelihood of rising rural indebtedness raise an important policy issue concerning the role of voluntary savings or thrift clubs in alleviating the problem of capital shortage among communal area farmers. The establishment of a network of viable savings clubs would help to reduce such indebtedness, particularly in the event of crop failure, as the farmers would purchase their inputs out of their own savings rather than through loans. The clubs could thus provide an effective vehicle for the mobilization of local capital resources for agricultural and rural 'development from within' at much lower unit costs both to the state and to the farmers. In the long run, and by broadening their range of activities, the clubs could even form the basis for the development of co-operative agricultural production units, in line with government policy for the socialist transformation of the economy and society.

3

A GHANAIAN RURAL COMMUNITY

indigenous responses to seasonal food supply cycles and the socio-environmental stresses of the 1980s

George J.S. Dei

INTRODUCTION

This case study examines the nature of the adaptive resources of the people of Ayirebi in south-eastern Ghana to local food supply cycles and to a national economic crisis of the 1980s which was triggered by world recession and aggravated by drought, bush fires and the influx of Ghanaian deportees from Nigeria. Ayirebi is a forest, food farming community with a population of about 4,300 made up of 2,021 males and 2,279 females.[1] It is located in the Eastern region of Ghana about 45 km from Akyem Oda and 180 km north of the Ghanaian capital, Accra. The town covers an area of approximately 1.75 km² The inhabitants are Twi-speaking, belonging to the matrilineal Akan sub-group known as the Akyem.

Ayirebi is without electricity or a piped water supply. Firewood is used for household cooking purposes. Kerosene lamps and lanterns supply household lighting. The water supply comes from streams, wells and rainfall. A non-bituminized feeder road from the nearest urban centre of Akyem Oda leads to Ayirebi and its surrounding villages. Government-built infrastructure in the town includes a police station, post office, health centre, two cocoa buying agencies and three public elementary schools. Community initiated facilities and services include a market square, public conveniences, a chief's palace and a rural bank. Other services which are prominent in the community are a private health clinic, three private vocational schools, and stores (for the sale of provisions, health drugs, toiletries, stationery, alcohol and drinks). There are also shops for artisans in blacksmithing, woodcarving, tailoring, hairdressing, as well as radio, watch, footwear and bicycle repair shops.

METHODOLOGY

This study was conducted primarily through active participant and non-participant observations of the everyday life in the community, the gardens and the bush, i.e. forest). The total study sample was 412 households, representing a quarter of all Ayirebi town households.[2] With the assistance of two local schoolteachers, these households were the focus of demographic, ecological and socioeconomic surveys which were conducted through interviews and the administration of questionnaires. The research sample had a total population of 1,543 people. Their age and sex distribution show 731 (47.4%) males; 812 (52.6%) females. The youth population, defined as 0–20 years, was 647 (41.9%) and the adults and elderly, 896 (58.1%). The active adult population (i.e. 21–64 years) is made up of 408 (45.5%) males and 488 (54.5%) females.

Four major climatological and ecological periods can be identified in the community. These are a main dry season that falls between January and March; a main rainy season from April to June; a pre-harvest season, July/August to September; and a main harvest season lasting from October to December. In gathering seasonal data on food supply cycles, the rainy season and the pre-harvest season were appropriately combined into the lean season. The seasonality of data dealing with household consumption and other economic activities which are presented in this study thus refers to a harvest season (October 1982 to December 1982); main dry season (January 1983 to March 1983); and lean season (April 1983 to September 1983). Each of these periods was sampled for data collection (between seven and fourteen days) to reflect seasonal differences and/or fluctuations in subsistence and other economic activities. From the initial 412 households, 20 representative households were further selected on a non-random, statistical basis for the more detailed study of the seasonal variation in household food procurement, distribution and consumption patterns, as well as to keep periodic activity diaries.[3]

THE LOCAL ECONOMY

Nearly 90 per cent of the adult population of this town depend directly on farming for their livelihood.[4] They produce both subsistence and cash crops. The local staples are plantains, manioc, maize, cocoyams, yams, rice and green leaf vegetables such as tomatoes, pepper, okras, onions, and garden eggs or egg plant. Other economic activities include hunting and gathering of wild forest resources, raising of livestock, and arts and crafts. Colonial and post-colonial changes introduced into the economy include the production of such cash crops as cocoa, kola nuts and palm oil, and a subsequent articulation with the national market economy and/or wage labour.

Since cocoa was introduced in 1879 as a commercial crop, it assumed an important role in the Ghanaian economy (Senyah 1984). The production of this crop was embraced by most farmers as it could easily co-exist with food farming for local consumption needs. Cocoa continues to be the chief crop of most major farming communities in the forest zones of southern Ghana. Its cultivation has dominated all other agricultural activities in the country (Wilks 1977: 490). The largest concentration of cocoa farms in Ghana today is found in the forest areas of the Brong Ahafo, Ashanti, and parts of the Central, Western, and Eastern regions of the country.

The leading position of cocoa in the farming economies of some Ghanaian communities has in the recent past suffered a gradual setback. Beginning in the late-1920s, a multiplicity of factors has diverted attention from the overwhelming production of cocoa to that of staple foods in certain forest farming communities.[5] One such community is Ayirebi, where the farmers have for a long time concentrated upon subsistence crop production.[6] Food farming there (and in the wider context of the Ghanaian forest zone) at the beginning of this century was mainly for subsistence, as there was no incentive to produce food crops on a large commercial scale. Owing to the bulky and perishable nature of most of the foodstuffs, and to the lack of communications and of proper organization among the producers, food farming in the past was usually not as profitable as cocoa. Therefore, most farmers preferred to grow cocoa wherever possible. But, as Boateng (1959: 68) points out, the scarcity of food which began during times of war and the need to feed large numbers of troops gave considerable stimulus to food production between 1939 and 1946. Government bulk-purchase organizations were set up at various centres; and the increasing demands of the urban centres over time have provided an incentive for food farming. It is therefore not surprising that, in all areas where the growing of cocoa has been unsuccessful, a food production industry has emerged. Contemporary events show that Ayirebi has become a town known for its food production. Food crops which have been grown there for local consumption are now sent regularly to the urban centres of Akyem Oda, Akyem Swedru, Koforidua and Accra.

Lineage control of land used to be a marked feature of the Ayirebi economy. Family land was vested in the matrilineage and every member of the lineage had the right to farm freely and to build on such land. The individual controlled only the usufruct of the land and could not alienate it nor transfer the rights to use family land or farms following matrilineal inheritance. Since colonial times, however, the emergence of such cash crops as cocoa and coffee, together with the demand for timber and gold mining concessions, have encouraged the wholesale alienation of land by some chiefs and lineage elders, at the expense of their subjects' rights, and a trend towards individual ownership of land has emerged (Dickson 1969; Manoukian 1964). In the contemporary period, the concept of family land is still upheld and a great proportion of food and non-food cash crops are produced on such land. However, land has come to be regarded more and more as a commodity, with the state and other wealthy individuals in the community striving endlessly to annex substantial portions of all land available. Thus the concepts of family land, stool land (land under custodianship of a chief), individually owned land and state land co-exist in the daily lives of the Ayirebi people (Dei 1987).

The 1982–3 survey of food farm plots which were owned by the sampled Ayirebi population showed an average farming household having between two and three separate farms. The study also showed household differentiation in the size of farm plots. Of the total of 412 research sample households, 124 (30%) households each had between 0–3 hectares of food farm plots; 140 households (34%), 3–6 hectares; 86 households (20.9%), 6–9 hectares; 41 households (10%), 9–12 hectares; and 21 households (5.1%), over 12 hectares (Dei 1987: 117ff.).

Generally, the basic production unit comprised a husband, wife, children and one or two matrilateral kin. This group could either be the sole production unit occupying a house or compound, or it could form part of a larger group occupying a compound. In this latter instance, the compound would be sub-divided into a number of separate (and independent) production units. It was also possible for a large group organized around a segment of the matrilineage (an elder woman and her sister or her daughters; or a man and his sister or his sister's children) to be the sole production unit within the compound or dwelling house (Dei 1986; Hill 1975).

Members of a household share farming activities, working together on the farm land that was acquired through the matrilineage and/or the custodian of the stool (i.e. village chief or sub-chief) or through the individual's personal effort (i.e. outright purchase) (Dei 1987). During periods of major economic activity (e.g. preparation of the land for farming, or harvesting of farm produce), when agricultural work intensifies, households may request the assistance of available extended family labour. Other 'external' sources of labour which are available to the household production unit include seasonal migrants from the northern parts of the country and casual wage labour provided by the town youth. There is also the formation of such partnerships as *nnoboa*, collective self-help groups of age mates helping each other in farming activities (Arhin 1983: 472). The adult males perform the task of clearing the forest and preparing the land for farming. Women, the young and the elderly in the household do the planting and harvesting of the crops, occasionally receiving some assistance from the adult males. Both sexes make joint efforts to keep weeds out of the food farms by using such farming tools as the hoe, digging stick, cutlass and axe. The principal methods of farming are shifting cultivation on bush farms, and intensive cultivation in the gardens and farm plots that are closer to the homesteads.

Ayirebi has fast become a market centre for the surrounding smaller communities. Traders and other visitors from both distant and nearby communities converge on the town to sell and buy foodstuffs as well as other goods at the local retail shops. There are established foodstuff dealers in the community, most of whom act as middlemen and women who make weekly bulk purchases of food items for traders in the Ghanaian urban centres. These foodstuffs are transported to the cities in hired articulated (trailer) trucks, which converge on the town on the Tuesday and Thursday market days. Established political structures, such as the Town Development Committee (TDC) and, more recently, the Committee for the Defence of the Revolution (CDR),[7] help to supervise the sale of farm produce in the local market, as well as to regulate the activities of the food dealers and urban traders.

Another sphere of Ayirebi's town involvement in the national market economy can be found in the government's employment of a handful of local residents as agricultural workers in the cocoa and oil palm plantations near Akyem Oda.[8] There is also rural migration of some town youth to the cities and urban centres, in search of non-existent white-collar jobs and a perceived 'better' standard of living.

The Ayirebi seasonal economic cycle (see Table 3.1) shows an alternation between periods of abundance and relative scarcity in household food supplies. Notwithstanding the effectiveness of local farming strategies in meeting the basic food requirements of households, periodic food shortages of brief duration do occur among households, particularly in the lean season of late May to August. Ordinarily, the situation may be attributed to unpredictable ecological and logistical factors, such as erratic and inadequate rainfall, crop destruction by pests and diseases or a temporary inefficiency in household productive strategies (e.g. poor management and/or sudden cutbacks in household labour supply resulting from illness during a farming season).

In 1982 and 1983 certain socioeconomic and environmental changes had implications for the successful adaptation of the Ayirebi community. The world recession adversely affected the Ghanaian economy, with the result that the national government could not afford to pay for imports of food, fuel, medical supplies, and other services and facilities which are associated with housing, clothing, education, transportation and communication, employment and industry. Consequently, local shops and stores were empty of such imported food items as milk and milk products, vegetable oil, flour, oats, sugar, tinned meat and fish. There were nation-wide shortages of basic amenities such as toiletries, textiles, health drugs and stationery. This period also coincided with the worst national drought in forty-eight years. Poor rainfall led to a general water scarcity, forcing people in towns and villages to walk long distances in search of water for household use and consumption.

Ayirebi town experienced these hardships, recording its poorest ever rainfall as reflected in both the total amounts and number of days of rainfall. In the past, Ayirebi annual rainfall had averaged 1,680 mm. Between October 1982 and September 1983, however, the total annual rainfall dropped to 933 mm.[9] The situation was more acute in the months from November to March when food planting activities reached a peak. Rainfed cultivation was affected. The drought also encouraged a series of bush fires that destroyed food crops, such as plantain, yams, cocoyams and cassava, as well as cocoa farms (Dei 1986). In the national food economy, the drought and bush fires worked in particular to depress production and to raise demand, which invariably worsened the problem of food availability. Even rural communities, which were spared the full impact of the world recession (on account of their minimal connection with the

63

Table 3.1 Seasonal economic cycle of Ayirebi, Ghana

Period	Ecological conditions	Productive cycle farming activity	Food cycle (food supply)
January–March	Main dry season: characterized by a dry parching land wind	Beginning of the agricultural season: preparation of fields, cutting, burning and tilling of farm land.	Farm food supplies partially available from previous harvest season. Cassava and cocoyams left in the fields may be harvested at this time. A second maize crop planted in early September is harvested February–March.
April–June	Main rainy season: with peak of rains in June	Farm activities continue with the sowing of seeds and crops at the onset of the first rains.	Food supplies in some households may be near exhaustion, causing episodic food shortages beginning in late May. The relatively lean season for cultivated foodstuffs continues to August, resulting in a marked dependence on forest plants and game. Early maize may be ready for harvest towards the end of August, helping to alleviate food shortage. Individuals may also make craft items to obtain additional income for food purchases at the local market.
July/August/September	Pre-harvest season: low rainfall in July–August and a second rainy season in September–October	Farming labour increases, including weeding of fields and tending to crops, thus increasing the expenditures of labour on the farms.	(see above)
October–December	Main harvest season: rains that began in September continue into October; November sees the beginning of the main dry season, continuing through December–March.	October–November is a period of intense harvesting of foodstuffs; the second half of the season is a period of leisure from farm work. Social gatherings and other community festivities peak. Crafts and collection of some forest products are undertaken but their contribution to household income and food supply is not particularly significant.	A season of relative plenty and abundance of food supplies for the community at large. The harvest permits the lavish consumption of food that accompanies the celebration of socio-religious activities. Brisk market activity as articulated trucks converge on the town to cart away foodstuffs to urban centres.

See also Fortes and Fortes (1936) and Hunter (1967).

global economy), were struck by the drought. For the Ayirebi community, an additional stress was the return of 298 town residents (about 7% of the town's population) who had been deported from Nigeria in the early months of 1983.[10]

HOUSEHOLD-LEVEL COPING STRATEGIES

The farming economy

The successful adaptation of the Ayirebi people to such stresses, particularly seasonal food supply cycles, can in part be attributed to the strength and viability of the household subsistence farming economy. There is great diversity of cropping patterns among local farmers. As part of indigenous safeguards against food scarcity and prolonged periods of hunger, local farmers have devised two growing seasons for maize. The main season maize, which is planted in April, may be ready for harvest towards the end of August (i.e. during the lean season), while the second crop, which is planted in early September, helps to extend the household food supply with the harvest in February and March. Cassava (*Manihot esculenta, Crantz*), the most versatile of the local staples, also assumed greater importance in the local diet during the stress period. Its capacity to grow and yield well on low fertility soils, its ability to withstand locust attack and drought, and its low cost of production provided the economic incentive for local farmers to use it as a replacement for staples such as yams. Cassava is a reliable year-round source of calories and a famine reserve crop. Many farmers leave unharvested actual surpluses of the crop on their farms, in order to constitute a valuable reserve in the event of a relative failure of other staples in a farming season. Its roots survive well when stored in the ground, to be harvested only when required. With its ability to increase the caloric yields and to extend the harvest season, cassava, in effect like cocoyam, serves to lessen the full impact of the lean season on most farming households (see also Annegers 1973: 256).

Like cassava, cocoyam (*Xanthosoma saggittifolium* or *Xanthosoma maffaffa*; and *Colocassia antiquorum schott* or *Colocassia esculenta*) is another important root crop that keeps well either in the field or in storage, and is therefore available throughout the year. The matured crop can be stored underground, as it were, and harvested only a little at a time when cocoyams are needed. In this manner it is common for harvesting to extend over four to five years (Karikari 1971: 9). An early and late crop of cocoyams may be taken, depending on the type of planting material. When land on which cocoyam has been planted is cleared again after a long fallow, small seedlings of the crop, with no traces of the tubers, may appear. These mature quickly and may be harvested after six months. Cocoyam is usually the first crop to be taken on land after clearing a secondary forest. Not only does the root crop spring up underneath burnt cocoa farms but its tubers can also lie dormant for several years. The local farmers have also developed a habit of digging out and removing mature cocoyams from the parent plant before the smaller ones are fully mature and the entire plant is lifted. By harvesting individual cocoyams when they mature, losses from rotting and premature germination are held to a minimum. This practice constitutes an important safeguard against future food scarcity or hunger.

As part of the indigenous strategies to cope with the stress on household food supplies, Ayirebi women devised additional processing methods for cassava and cocoyam to make them edible irrespective of their natural state. The skin of cocoyams under normal favourable circumstances would be discarded. However, during this period of hardship, the women spent some time drying the cocoyam skins before milling them into a flour and preparing them at extremely hot temperatures for consumption.

Another coping strategy was the pragmatic dependence on the local market on the part of households. Particularly during the lean season, households facing deficits in their food supply would obtain their food requirements from the relatively well-off farmers who had been able to produce surplus for the market. Payments for such market foods varied from monetary payments to exchanges of forest collected food items.[11] It is during this lean season that some degree of difference emerges between households and their diets in terms of both content and source of food supplies. Distinguishing Ayirebi households by income, the picture that emerges suggests that differences in source and content of household diets can in part be attributed to differences in economic status. Table 3.2 shows the responses by income status of twenty representative households which were selected non-randomly for the study of the sources of the household food supply during the lean season (see also Dei 1986).[12] The income brackets were chosen arbitrarily.[13] In meeting the cash demands of augmenting household food supply from the local market, households rely on accumulated earnings from previous sales of farm produce, wild forest resources, sales of arts and crafts, wage labour, services and rental income, remittances and welfare, and livestock rearing.

Exploitation of wild resources

Hunting and gathering has been an economic strategy which was utilized in the past by some Ayirebi households in response to the seasonal fluctuations in food supplies. In 1982 and 1983, households were observed to be reverting to dietary patterns that relied heavily upon the surrounding natural environment. A wide variety of edible and non-edible (medicinal) wild products, such as roots, fibres, leaves, bark, fruits, seeds, nuts, insects, molluscs, sap and syrup was exploited largely by Ayirebi women and children to satisfy basic household needs. The women experimented with four new varieties of plants for cultivation purposes. These were: *Dioscorea praehensilis* (bush or forest yam), a root crop locally referred to as '*ahabayere*' and normally classified as a semi-wild yam; *Afzelia bella*, a wild plant locally referred to as '*papaonua*', the leaves of which are used as vegetables in soup; *Napoleona vogelii*, known as '*obua*', whose fruits are eaten as a delicacy; and *Blighia welwitschii*, known locally as '*akyekobiri*', the leaves of which are also used to flavour soup.

Table 3.2   Food strategy responses of twenty Ayirebi households to seasonal nutritional stress by income status

Status*	Contribution to household food supply from four sources in the lean (pre-harvest) season (%)			
	Farm	Bush	Market/Store	Other (kin and close friends)
Wealthy n = 3	75	2	16	7
Middle n = 10	65	8	19	8
Poor n = 7	55	20	10	15

*Wealthy = annual cash income (i.e. market proceeds of all economic production) of C 8,000 + ;
Middle = annual cash income between C 4,000 and C 8,000;
Poor = annual cash income of less than C 4,000;
where one cedi (C) = US$ 0.36 (1982–3).

Households found local substitutes for imported sugar and cooking fat in honey and palm oil respectively. After extracting the oil from the palm kernels, the women utilized some of the palm oil products, together with wood ash, to make local substitutes for the scarce imported soap. The hunting and trapping activities of Ayirebi farmers and male professional hunters also provided the community with a steady supply of bush animal protein. These included: mammals (grasscutter, *Thyromomys swinderianus*; giant rat, *Critetomyx gambianus*; antelope, *Neotragus pygamaeus*; brush-tailed porcupine, *Atherurus africanus*); reptiles and aquatic species (such as molluscs, crabs, clams, tortoise, giant forest snails); and a wide variety of birds, invertebrates and insects (Dei 1986).

Non-agricultural production

The research period also witnessed a marked resurgence in local skills in non-agricultural production, particularly among the male popula-tion. The Ayirebi environment provided enough raw materials to produce a wide variety of products for both household use and sale on the market. Activities included blacksmithing, tinsmithing, basket-weaving, straw-work, barkcloth-making, woodcarving, mud-brick and tile-making. These are viable economic activities that enable the practitioners to supplement their household income and food supply.[14] Households which were unable to afford the services of skilled artisans or the high prices of imported building materials (e.g. cement, iron roofing sheets) made some cash savings by constructing and repairing their houses themselves out of local raw materials. During the research period, when imported cement and iron roofing sheets were scarce, most houses were constructed or repaired in the community with mud-bricks and thatch roofs. Both basketry and thatch roofing manufacture are also crucial for food storage purposes. The resurgence in the importance of blacksmiths in the community can be attributed to the absence from the local market of such farm implements as cutlasses, knives, and axes. Local blacksmiths are increasingly relied upon to produce these implements. They accept old farm implements from the local farmers and reforge them for recycling (see also Posnansky 1980: 2419). This revival in the work of blacksmiths is confirmed by the interest shown by some local male youths in becoming apprentices of the trade.

Remittances and welfare

A feature of wage labour of some significance to the household economy and to rural adaptation in general is the remittances of urban wage earners to their kin and friends in the village. Within the Ayirebi community the wage earners are predominantly males and their remittances usually take several forms. These include clothing, footwear (both old and new), money (to help kinsfolk in their farming, or as kin contribution towards a family project, funeral celebration or household maintenance), and imported consumer goods such as food items and medicine. Cash remittances, in particular, are significant in the household farming economy. The money received can be used to hire seasonal or casual wage labourers on the family farms.[15] Studies of money order transactions at the Ayirebi Post Office during the three-year period 1981–3 show a relatively steep rise in cash remittances in the difficult months of May and June 1983 (i.e. lean season period) over previous years (see Table 3.3).[16] A possible explanation for this is that, during the crisis period, when urban wages were inadequate to meet rising food prices, urban wage earners found it necessary to devise alternative strategies. They remitted cash to their rural kinsfolk very frequently, in order to obtain some food items directly and also to assist kinsfolk in hiring seasonal or casual wage labour for family farms. In some cases, the cash remitted was clearly earmarked for the establishment of new farms on the family plot on the wage earner's behalf.

COMMUNITY-LEVEL COPING STRATEGIES

Self-help and mutual aid have been cornerstones of the Ayirebi community-wide responses to the hardships of the early 1980s. The remarkable extent of social responsibility that was exhibited by the people is a testimony to the effective organization of the Ayirebi social and economic structures as well as its political leadership patterns. For example, to cope with the problem of scarcity of good drinking water during the 1982 drought, the eight-member Town Development Committee (TDC) in conjunction with the local Committee for the Defence of the Revolution (CDR) effectively mobilized the Ayirebi male community to construct both compound and community wells.[17] With regard to bush fires, a number of small fire-fighting groups of farmers who shared boundaries was formed. Town residents of both sexes, through a labour pool, assisted in the replanting of individual burnt farms. Victims of bush fires were financially assisted through voluntary contributions from the community.

The reintegration of the 298 returnees from Nigeria into Ayirebi society was viewed as the responsibility not only of the individual

Table 3.3 Monthly breakdown of the yearly record of money order transactions conducted at the Ayirebi Post Office (1981–3)

Year	JAN		FEB		MAR		APRIL		MAY		JUNE		JULY		AUG		SEPT		OCT		NOV		DEC		TOTAL	
	N^1	A^2	N	A	N	A	N	A	N	A	N	A	N	A	N	A	N	A	N	A	N	A	N	A	N	A
1981	12	850	4	205	—	—	1	22	—	—	2	34	1	26	4	123	7	202	8	350	3	125	9	518	51	2,455
1982	18	1,307	9	620	2	32	1	29	5	75	2	326	2	62	5	240	8	407	9	353	8	238	10	620	85	4,309
1983	23	1,876	10	877	2	150	2	53	11	850	15	1,504	6	408	7	210				*Not available*			—	—	76	5,928

1 N = Number of individuals reporting to the Post Office to conduct money order transactions.
2 A = Total amount involved in all transactions for the particular month. Amount in cedis (C), where one cedi = US$ 0.36 (1982–3).

household and families concerned but also of the wider community. The local chief, after consultations with his elders, took the first initiative and released a total of 2.5 hectares of stool land to the returnees for farming on a co-operative basis. Town residents with uncultivated lands were also encouraged to lease them out to interested individual returnees for farming. To reciprocate such community gestures, some returnees shared part of their acquired wealth from Nigeria with other town residents.[18] They also formed a voluntary body, the 'Ayirebi Agege Boys and Girls Association', which occasionally conducted two-hour clean-up exercises to improve sanitary conditions in the town. They assisted the various town committees to mobilize the local population to conduct other self-help projects effectively (e.g. the construction of four new public conveniences, using primarily local materials, within a period of two months between April and May 1983).

As part of the community coping strategies for the stress in the domestic food economy, town residents communally brought additional land into production. A series of community level co-operative farms were established (e.g. Town Co-operative farms, farms of the returnees from Nigeria, and the Ayirebi community farms). The local CDR in 1983 also secured an additional two hectares of stool land for community farming. During the same period, a group of women in the community obtained four hectares of stool land in order to establish a communal rice farm. In May 1983 the local chief, attempting to lessen the effects of the lean season's food supply on all Ayirebi households, appealed to the successful farmers to help their unfortunate counterparts who were experiencing hardships from poor harvests. Farmers with surplus food for sale were to make sure that local needs were satisfied first before supplying external markets. The appeal demanded that foodstuffs for sale were to be displayed at the Ayirebi market, in order for needy households to have the first right of purchase, before being sent to the established food dealers in the community. The local CDR saw to the execution of this measure which, despite a few grumblings, was largely adhered to by the farmers.

The extent to which the varied coping strategies were gender specific has already been pointed out: men hunted, engaged in non-agricultural production, constructed household and community wells and performed wage labour, while women gathered wild products, experimented with the cultivation of varieties of wild plants, made local substitutes for imported soap and vegetable oil, and developed

71

Table 3.4 Summary results of a seasonal weekly survey of the portion of a 12-hour day spent on various activities by both sexes in twenty representative Ayirebi households (no. of hours)

Seasonal cycle	Farm work[1]		Food processing[2]		Transport and maintenance[3]		Building repairs[4]		Child care[5]		Non-agri. production[6]		Total work	
	M	F	M	F	M	F	M	F	M	F	M	F	M	F
Post harvest season (Jan–Mar)	5.1	3.0	0.3	3.5	0.5	0.8	0.5	—	0.1	1.0	0.3	0.4	6.8	8.7
Lean season (April–Aug)	4.0	3.7	0.4	3.0	0.4	0.9	0.9	0.1	0.2	1.0	1.1	0.5	7.0	9.2
Harvest season (Oct–Dec)	1.9	2.2	0.6	3.7	0.7	1.3	0.6	0.1	0.1	0.8	1.4	0.7	5.3	8.8

1 Refers to actual work relating to field preparation, cutting, planting, weeding, harvesting and transporting produce to the household.
2 Activities associated with food processing and preparation, e.g. sun-drying, cleaning, peeling, winnowing, milling, cooking and serving of meals.
3 Includes collection of firewood and other fuel, fetching of water, laundry, sewing and household cleaning, washing of dishes, transportation of foodstuffs to market, and marketing.
4 Construction of new homes, repairs on existing structures.
5 Tending of children to bed and school; child learning; health needs of children.
6 Time devoted to wood and other fibre work.

additional processing methods for some crops. A seasonal weekly survey, which was based on observations and on interviews regarding household daily activities and the time spent on each, was conducted among adults of both sexes in the twenty representative households mentioned earlier. Data are summarized in Table 3.4. A basic finding is that men's work, although at times more strenuous, is generally limited to specific seasons unlike that of women (see also Brandtzaeg 1982; White 1976). The men may spend a great deal of their time performing farm work, but the contribution of women is equally significant. The role of women in agricultural production receives added significance when work relating to travel and household maintenance, food processing and preparation for consumption is taken into account. Men do not take part in some of the activities relating to household maintenance but, when all groups of activities performed in a day are totalled, the labour input by women, measured by time spent on the various activities, is far greater than that of the men. However, it should be stressed that, on the whole, the research period witnessed a remarkable degree of co-operation between the genders to find solutions to common problems both within the household and community and on the farms.

Normally, one would expect that the effects of severe economic hardships would be manifested in the conduct of social relations. That is, social conventions such as 'sharing, hospitality and generosity will go by the board in times of scarcity', while individual household needs and welfare become the prime concern (see Turton 1977: 188, on Ethiopia). This was not the case in the Ayirebi community. Largely through self-help, mutual aid, sharing and generosity, the local population was able to cope with ensuing hardships. There was a cultural awareness and acceptance that the main causes of the environmental crisis were human induced. The crisis was widely attributed to the breakdown of respect for customs, including one's obligations to kin and neighbours. There had been frequent condemnation from the aged in the community of the contemporary rising trend towards greed and the craze for self-aggrandizement at the expense of others. There was also a feeling that certain customary restrictions that were being increasingly ignored (e.g. restrictions on bush burning during certain periods and under certain conditions) were based on practical knowledge which had been acquired by the town ancestors. Such knowledge derived from long-term experience of subsistence strategies in an environment which was subject to the vagaries of climate (Posnansky 1984: 2163). In large measure, then, the drought was seen as a punishment from the gods and ancestors in order to make the living aware of their neglect of adaptive customary behaviour and of the need to live harmoniously in the community (see also Turnbull 1972: 284). It is in this light that one has to understand the success of the local authorities in mobilizing the people to combat the hardships that they were facing and to fight for group rather than individual survival.

The study of criminal records at the local police station does not provide any significant evidence of strain in social relations during this period. The 1982–3 police records on crime, and other household or community disputes, show no marked increases over previous years. Sixty-one cases, including bodily assault (fighting), drinking abuse and insulting behaviour, breaking of national curfew hours, theft of household property and food crops (including bush animal protein from farm traps), were reported to and investigated by the local police officers.[19] In living memory, there has been no charge of murder, manslaughter or sexual abuse in the community. Those cases of bodily assault that have been reported are usually in the form of fist-fighting; implements such as knives and cutlasses are rarely used to inflict wounds. Of 15 bodily assault cases which were reported during the research period, knives were used in only four to inflict injury. The foregoing, however, is not intended to mean that Ayirebi social life was without an incident which could be attributed to the hardships of the stress periods. During the lean months of May to June 1983, there were occasional complaints of theft from farmlands involving food-crops such as cassava, plantains and cocoyams. These incidents were rarely reported to the police. The local chief, through the town crier, communicated to his people the need to be vigilant and he warned that offenders, when caught, would be severely dealt with. Other indicators of the effects of economic stress include a few observations of people calling in outstanding debts and pressing claims which, in less difficult or normal times, they might not have done (Turton 1977: 190). On the whole, community life was peaceful and in stark contrast to the stories of increased crime rate and commotion in the cities and urban centres, as reported in the national daily newspapers.

DISCUSSION

This study has focused on how one Ghanaian community has responded to the processes of change. The vitality of the rural survival strategies of the people of Ayirebi in combating the effects of seasonal food supply cycles and of other contemporary socioeconomic and environmental stresses has been explored. It was shown how local self-sufficiency in food and other basic requirements was achieved in the community through four related factors: the strength and viability of the subsistence farming economy, which showed great diversity in cropping pattern; the adoption of hitherto underutilized subsistence practices (e.g. hunting and gathering of wild resources, and a resurgence of non-agricultural production); a pragmatic dependence on the local markets and/or cash economy; and the extent of social responsibility which was exhibited by community members.

While the coping strategies which are discussed here may not be altogether new, they have re-emerged in the contemporary scene in ways that clearly demonstrate the creativity and resourcefulness of the local people. In the Ayirebi community the local people are using all of the resources that are available to them from past experience to combat the stresses and strains that are brought to bear on existing means of satisfying basic necessities of life. Such micro-level studies of the nature of the contemporary adaptation of rural communities provide useful lessons not only in crisis containment but also, more importantly, in development planning; specifically, on the necessity of building thriving, self-reliant, self-sustainable local and regional communities in contemporary Africa.

This study raises issues regarding the nature of local–State relationships, particularly in the context of some of the structural adjustment programmes that are recommended by the World Bank for tropical African economies (WB 1981, 1984a). The active role of the State in promoting export-led development through cash crop production and the free play of market forces is seen as essential for the long term development of the African economies. Primary importance is attached to external assistance in the developmental process (Hinderink and Sterkenburg 1987: 271–3). A basic problem with export-led development relates to the continued subordinate position of African economies in international terms of trade and division of labour. The effects of the world recession of the 1980s on African economies reaffirm the fact that the State cannot guarantee the availability and growth of fair markets for national export products. This means that the State has to reassess its role in the developmental process. It must promote reforms that not only make for the efficient allocation of available resources, but also utilize indigenous creativity and resourcefulness in order to improve upon the living conditions of the rural majority.

The Ayirebi case study offers some useful insights in this direction. A socially responsible local leadership is leading the way to addressing pressing issues of communal interests during stressful situations. There is grass-root level participation in village decision-making, and the crucial issue of access to productive resources, particularly land, is being addressed. Local checks on the activities of 'external' forces in the domestic economy (e.g. marketing intermediaries; food dealers) ensure that external demands do not adversely affect local needs and requirements. Other areas of spontaneous action on the part of Ayirebi farming households to combat the socio-environmental stresses on the local economy have been spelt out. The noted inter- and intra-village integration involving households' pooling of resources and labour power constitute effective strategies for self-sustained and self-reliant development.

Participation in the cash economy can, to some extent, make a contribution to a farming household's ability to cope with national economic crisis, by making other sources of revenue available for other household purchases. However, a deeper involvement at the expense of subsistence production can also have detrimental effects in terms of local food self-sufficiency. The consequences of cash tree cropping on the local food requirements of developing African countries have been far reaching. In most of these countries, the redirection of agricultural efforts away from food crop production to cash tree crops has worked to the detriment of household, community, and national levels of food adequacy. The question which readily comes to mind is whether the observed local food self-sufficiency in the Ayirebi community is attributed to the primary importance of food production, for both household consumption and the external market, over non-food cash cropping (e.g. cocoa). There is limited and inconclusive evidence to answer this question, and further research in other Ghanaian communities is needed. At the national level, however, certain facts are clear and relevant. Since the turn of this century, a significant factor influencing food production in Ghana has been the introduction of cocoa as a cash crop. In areas where cocoa is dominant, the best land is allocated to this crop, while food crops are relegated to the poorer land. The continuing government emphasis on export cropping to the neglect of small-scale rural food production has necessitated the importation of food items to supplement local production. In fact, some of the more recent national policies to combat the effects of the 1982–3 drought and bush fires (such as those relating to cocoa rehabilitation and replanting) involve large tracts

76

of potential foodcrop land, which may seriously influence total food output in the country in some years to come.

Further research is also needed to assess the degree to which the coping strategies of the Ayirebi community have been continued long after the 1982–3 drought. But it is important to point out that, after the rains came in June 1983 and in 1984, Ayirebi households could still be observed as being engaged in the major survival strategies discussed. These included: a continuation of food collecting and hunting activities; experimenting with additional varieties of plants for cultivation purposes; non-agricultural production; use of local substitutes for imported soap, vegetable oil and sugar; and expanded processing methods for staples such as cocoyam and cassava. Future research is needed to identify whether or not there is a continuance of the changes that were observed in land use and allocation, in food crop diversification, in farm labour adjustments, in household remittances and welfare, and in the wider community responses to ecological and other stressors.

Further studies are necessary before generalizations can be made from the results of research in Ayirebi. Nevertheless, this study questions conclusions about the state of African economies which are based solely on macroeconomic data, themselves measuring for the most part production for the external market and omitting from consideration local survival strategies.

ACKNOWLEDGEMENTS

This study was conducted in 1982 and 1983 for my doctoral dissertation in Anthropology at the University of Toronto. I am grateful for the comments received on this study from Professors Richard Lee and Maxine Kleindienst of the University of Toronto, Professor Merrick Posnansky of the University of California, Los Angeles, and Professor Alejandro Rojas of the Faculty of Environmental Studies, York University, Toronto. I am deeply indebted to the chief and people of Ayirebi, near Akyem Oda in southeastern Ghana, for their warm hospitality. Funding for this project was provided by the University of Toronto through the award of a Connaught Scholarship.

NOTES

1 This figure is based on projections of Ghana's population growth for the 1980s from the 1970 census of 3,450 for the Ayirebi town (Central Bureau of Statistics 1982). The official report of a late-1984 population census which was carried out in the country had not been published at the time of writing.

2 The household refers to a group of people usually (but not necessarily) living in a house or compound who have a common food supply, pool their incomes for common support and regularly use and share the contents of a cooking pot.

3 Food records were kept in both the local language, Twi, and in English. In some of the households, the help of those with some formal education was sought for this detailed study.

4 A study of the occupational characteristics of the 412 sampled household heads (of whom 96 [23.3%] were female and 316 [76.7%] were male) shows that 396 (96%) consider farming their primary occupation while two mentioned hunting and trapping. The remaining 14 primary occupations were: herbalist (1); fetish priest (1); tailor and seamstress (2); trader (4); police officer (2); school teacher (2); shoe repairman (1) and transport driver (1). Fifty-six (13.6%) of the household heads indicated hunting and trapping as a secondary occupation.

5 Among the factors accounting for this development are ecological problems (drought, bush fires, soil erosion), diseases (such as swollen shoot), the problem of agricultural labour, and the failure of successive Ghanaian governments to provide adequate incentives to cocoa farmers (see Senyah 1984). Because of the declining revenues of some cocoa farmers, individual attempts are being made now to devote land to the cultivation of foodstuffs for both domestic and commercial purposes (Atsu 1984).

6 Hill (1963) has detailed the crucial role played by a group of migrants from Akuapem, Ga, and Shai in the hills above Accra in creating the Ghanaian cocoa industry. Starting in 1892, they migrated eastward to the present state of Akyem Abuakwa, where they bought large tracts of uncultivated land from absentee landlords. These migrants occupied an area further to the south east of Ghana than where Ayirebi town is located, and their settlements post-date the known early history and evolution of the Ayirebi community. The early settlers of Ayirebi were food producers. In later years, when the successful cocoa enterprises of the Akuapem migrants became generally known, cocoa growing became a feature in the Ayirebi economy. However, the cocoa farmers of Ayirebi (unlike those migrants of Akuapem) primarily relied on lineage and stool land (land under custodianship of a chief), as well as family labour, while combining their cocoa farming activities with food production (Wilks 1977: 526).

7 The members of the Town Development Committee (TDC) are selected by the local chief in consultation with his elders. Membership of the Committee for the Defence of the Revolution (CDR) is open to all town residents. This latter body was formerly called the People's Defence

Committee (PDC). It is a more recent political action group, the idea for which was introduced by the ruling Provisional National Defence Council (PNDC) government since 1982.

8 This wage sector is, however, not a well developed feature in the local economy. Fewer than a hundred individuals in the town are employed by the government as agricultural workers. Interestingly enough, although these are supposedly full-time employees collecting monthly salaries, they normally report for duty only twice a week and spend most of the remaining time working on their farms. In my conversations with agricultural workers, they unanimously pointed out that the government rarely pays their wages regularly, and that at times they are not paid at all for three or four successive months. The workers are thus forced to embark upon their own farming as a means of obtaining a sustained livelihood.

9 See Ghana Meteorological Services, Annual Rainfall Reports, Headquarters, Accra, Ghana.

10 The 298 returnees were made up of 210 males and 88 females with ages ranging from a three-month-old baby to a 49-year-old adult. Two hundred and seventeen of the returnees resettled in Ayirebi between January and April 1983, with the other 81 arriving between May and October 1983.

11 The importance of the local market increased during the season of relative scarcity in so far as household food consumption needs are concerned. The market prices of the basic staples (cocoyams, plantain, cassava and yams), as well as other protein-source foods (such as meat and fish), also tend to fluctuate with seasonal variations in supply. In the lean season, because of relative scarcity of some basic staples on the market, higher prices are demanded for the products on sale. But, just as market purchases of basic staples for local household consumption may rise during this lean period, those of meat and fish, as well as other protein-source foods, show a steady decline. The lean season witnesses an increase in the relative contributions of game from hunting and trapping activities, which supply most of the protein required in household diets. With the emergence of the harvest season, the importance of the market for household food supply reverses, with a period of increasing dependence on self-produced foodstuffs and a decrease in the contribution of purchased basic staples to the calorie content of the average household diet. However, as this period also coincides with the festive and ceremonial season which entails increased consumption of protein-source foods, market purchases of meat and fish (particularly of smoked herring and catfish) show a steady rise over those of the previous season.

12 In providing such information it is borne in mind that monetary income alone may not always give a complete and accurate picture of the economic status of the individuals involved. The income data should be supported by additional consideration of other important assets in the form of such immovable property as house and land. Before drawing conclusions on economic status, this researcher compared the income data and his observations of household property (e.g. land, houses) with the views of other community members. The income data are considered to be a relatively good index of, or a close approximation to, economic status.

79

13 The minimum daily wage was C 12 at the beginning of the research period. Although incomes expressed here may appear rather high for the predominantly small-scale rural farming households, they are if anything on the low side. Owing to prevailing high inflation levels in the country (in 1982–3 over 350%), these incomes do not reflect actual purchasing power.

14 There were a few instances, observed during the research period, of local artisans (blacksmiths, woodcarvers, bamboo and basket weavers) accepting food items for their manufactured products, as was usually the case in the past.

15 Most of the wage earners living outside the community, in order to make sure that their future food requests to rural kinsfolk will be heeded, try to remit some money to the farmers at the onset of the farming season and also during the harvest season. Other wage earners also provide some of their income to their rural kin so that the latter can employ farm labourers to set up local farms on behalf of the urban workers.

16 A certain amount of caution should be exercised in reading and interpreting the figures because of the problem of inaccurate record keeping at the Post Office. It is also possible that some monetary transactions of this nature might have been conducted by a few local residents at the major Post Office of Akyem Oda rather than at Ayirebi, or that people do not cash their money orders immediately. Also, most of such monetary remittances do not necessarily come to Ayirebi residents through the local Post Office. Substantial sums of money are received through the personal visits of the donors themselves, or via their friends and other visiting relatives. Some allowance must also be made for inflation and the increase in money supply in the country over the past two years. None the less, the comparative figures for three years lend credence to an argument for the increasing importance of cash remittances during the stress period of 1982–3.

17 The digging of wells was initially a voluntary exercise. As the water supply situation worsened and additional community wells had to be established, the local CDR instituted a penalty of C 20 (i.e. US$ 7.20 in 1982–3) for households whose male adults failed to provide communal labour.

18 The arrival of the returnees temporarily brought some goods into the community, such as soap and toiletries, clothing and textiles, tinned foods, health drugs, building materials (iron nails and tin roofing sheets), as well as household effects, such as tape recorders, stereo equipment, mattresses, trunks and suitcases. In addition, there were ten chain-saws, two mini-Toyota buses, and five electric generators, the presence of which helped in the farming, transportation, and energy requirements of the local population respectively.

19 During the fieldwork, countryside curfew hours were in force from 10:00 p.m. to 5:00 a.m. daily. Also, the figures for cases reported to the local police include crime and assault cases from other small communities within the immediate vicinity of Ayirebi. Civil cases involving land disputes and marital conflicts usually end up in the local chief's arbitration courts rather than with the police. Cases involving witchcraft

accusations are normally dealt with by the local fetish priests. Finally, it is stressed that not all criminal or assault cases necessarily end up in the hands of the police. Therefore, caution must be exercised in placing too much reliance on the figures provided.

4

THE CO-OPERATIVE CREDIT UNION MOVEMENT IN NORTH-WESTERN GHANA
Development agent or agent of incorporation?

Jacob Songsore

INTRODUCTION

Although it is generally accepted that credit plays a crucial role in the expansion and development of productive forces, social scientists, in addition to the World Bank and other donor agencies, are only now assessing the part that bank credit has played in the centralization and concentration of capital in the process of accumulation. This process had long been observed by Lenin in his analysis of finance capital (Lenin 1978: 30–59; see also H.T. Thomas 1988: 8).

At the micro level, the role of the financial system is to provide adequate savings and credit facilities to individual households whilst, in a macroeconomic context, financial institutions have the basic functions of channelling capital from savers to investors. An efficient financial system is consequently assumed to have a considerable positive effect on increasing welfare and stimulating economic activity and development. As a result 'credit is now the largest component in the World Bank's agricultural lending . . . and many governments of developing countries have assigned to credit the lead role in rural and industrial development programmes' (H.T. Thomas 1988: 10).

In the normal workings of a capitalist economy, financial institutions collect all kinds of financial revenue which they place at the disposal of the capitalist class. A study of the workings of formal financial institutions in the underdeveloped economy of rural Ghana has come up with similar conclusions. As the researchers put it:

> By providing liquidity in quite a selective way, primarily to the powerful and alert members of rural society, the banks

contribute greatly to a redistribution of wealth in favour of these 'elites'. Although the rural people would not use this terminology, it is the view held by the majority of the rural people and one which they do not hesitate to express; 'They only help the big shots to become richer'.

(Bentil *et al*. 1988: 126)

The *laissez-faire* operation of banking institutions in Third World countries and more especially in Ghana has had two consequences: first, the denuding of impoverished rural regions which are of little interest for the accumulation of much of their locally generated capital and, second, the creation of isolated pockets of rural poverty, i.e. the classical labour reserves. The latter have few modern financial institutions. Indeed, where these exist, they hardly serve the interest of the numerous illiterate rural peasant men and women in the lower ranks of the social ladder.

For example, in Ghana it is widely recognized that commercial banks have served to mop up and transfer rural savings to the capital markets of the Accra-Tema metropolitan area, the national capital, and to a lesser extent to other large metropolitan areas such as Kumasi and Sekondi-Takoradi. Such capital drained from cocoa and food producers has then been reallocated for foreign monopolies and other business interests which are located in the non-indigenous economy, i.e. those economic sectors with very few spin-off effects on the petty commodity sector of the national economy.

Of late, in order to stimulate a stagnating rural economy which has grave consequences for the national economy, rural banks have been established as a way of reducing these flows. This does not imply that there have been no initiatives on the part of the rural poor to adopt mitigating or survival strategies as a way of coping with the haemorrhage of capital and the limitations that are imposed on their productive activities and welfare services by their incorporation into the capitalist world economy.

These initiatives have taken the form of rotating credit associations known as '*susu*' and the development of the co-operative credit union movement. These indigenous rotating credit associations provided the basis for the development of the new organizational form of co-operation through the credit unions, since, without this spirit, the credit union movement might never have gained ground in the region. Where such organizational forms to defend the interest of the poor are lacking, both the rural and urban poor fall into the hands of 'usury' money lenders who step in to fill the gaps that are left by finance capital.

North-Western Ghana, as the name implies, lies in the north-western corner of Ghana. For the purpose of credit union administration, North-Western Ghana includes the present Upper-West Region, and the Bole and Damongo Districts within the Northern Region which currently operate under the Wa Chapter Head Office of the credit union movement. The region represents the worst pocket of extreme poverty in Ghana, with no modern industry except a cotton ginnery, few parastatals and a predominance of peasant agriculture using the most rudimentary technology. Out-migration to the southern mines and cocoa farms has been the lot of young men, leaving behind a population comprised largely of women, children and the elderly (Songsore 1983; Songsore and Denkabe 1988).

The underdevelopment of productive forces in the region can best be understood in the context of internal contradictions arising out of its colonial incorporation into the world capitalist sphere as a regional periphery of a peripheral state. All available studies indicate that the poverty of its people has further deepened as a result of, on the one hand, the ripple effect of the contemporary crisis of underdeveloped capitalism and, on the other, the marginality of the 'distant' poor in the area from the focus of the ongoing Economic Recovery/Structural Adjustment Programme (ERP/SAP) of the state. The politically powerless peasant producers in the region, whilst sharing few of the benefits of the ERP/SAP, have had many of the burdens of economic recovery pushed on to them.

Since this IMF/World Bank fashioned programme is based on an export-led strategy of economic recovery, resource flows in terms of loans have been directed to cocoa producers, and to the mining and timber industries in southern Ghana, to the neglect of food producers in the region. By contrast, as a result of the effects of devaluation of the local currency, liberalization and price de-regulation on the prices of inputs and consumer goods, the interregional barter terms of trade have shifted against rural food producers in the region. Utilization of health, potable water and educational services have all declined because of the imposition of high user charges. The imposition of higher user fees and the withdrawal of supplementary feeding programmes have caused a sharp drop in attendance at child and maternal welfare clinics. This is occurring at a time of drastic cutback in so-called nonproductive expenditure in the social sector. The worst hit have been women and children from very poor households.

Among these, children under five and pregnant and lactating mothers have been most affected. The overall quality of life among these people is the worst in Ghana (Songsore 1989).

Credit, by enabling peasants to expand and develop income-generating activities and by supporting payments of water tariffs, health charges and school fees, could be a vital tool for empowering the rural poor to cope with the consequences of the current crisis. By so doing, credit from these local co-operative credit unions could serve as vital instruments for achieving development from within and below (Mackenzie and Taylor 1989: 133).

In response partly to the area's neglect since the colonial period and partly to the pressures for cash mediated subsistence consumption and the intensification of commodity production, the Co-operative Credit Union movement in Ghana was born in the north-west of Ghana in 1955 (Songsore 1982: 3). The movement was to serve as a focal point for the endogenous, grass-roots mobilization of local resources for local development within the limits set by the national and inter-national system. It is very often the belief that, through the organized collective action of communities at the local level, the poor can fashion a suitable response to their progressive marginalization by the normal workings of finance capital.

This paper reviews state policies and the development of institu-tional credit in this largely neglected rural region and attempts to situate credit unions in the context of this general development. A major proposition which is advanced is that the co-operative credit union movement has served more as a vehicle for the penetration of capitalist production relations in this regional periphery rather than as a development agent addressing the needs of the majority of the members.

Pertaining to the above proposition, the paper discusses urban–rural contrasts in access to credit; differential class and gender access to credit; and peasant protest and crisis in the credit movement. In conclusion, suggestions are made which are aimed at recapitalizing and redirecting the credit unions towards the genuine developmental aspirations of the rank and file membership, especially women who are among the least heard. It is only through such reforms that the movement can serve the interest of the poor and thus become an effective coping mechanism for the poor.

THE POST-COLONIAL STATE AND THE DEVELOPMENT OF INSTITUTIONAL CREDIT IN NORTH-WESTERN GHANA

As a result of the low level of development of productive forces, a situation which was bequeathed by colonialism (Bentil *et al.* 1988: 11; Songsore 1983), both foreign and national banking institutions have been slow to develop in North-Western Ghana in contrast to their early development in the more monetized export-oriented areas of Southern Ghana. In the colonial period, for example, only one branch of the Standard Bank was located at Wa to serve the whole region. It was later replaced by a branch of the Ghana Commercial Bank in the same town. This was the only source of bank credit in this area of over half a million people until the 1970s.

Two new branches of the Ghana Commercial Bank have now been opened in Lawra and Tumu. This is partly in response to the expansion of government administrative institutions and schools in these district centres but also because the rural credit unions, which have long existed in these districts operated deposit accounts with this Bank thereby ensuring their viability. In order to stimulate agricultural development in the area, the Agricultural Development Bank now operates branch offices at Wa, Lawra and Tumu. Besides these, there is a branch each of the Social Security Bank and Co-operative Bank at Wa which, since 1983, has served as regional capital of the new Upper-West Region comprising the districts of Wa, Lawra and Tumu. The Co-operative Bank also has branches at Tumu, Jirapa and Hamile, all in the Upper-West Region. Two rural banks now exist in Jirapa and Nandom through the energies of their Youth Associations.[1]

Similar developments in commercial banking have taken place in Bole and Damongo Districts which, although they lie within the sphere of the Upper-West Chapter of the Co-operative Credit Union Movement which is headquartered at Wa, are administratively within the Northern Region. Damongo and Bolé Towns each have a branch of the Ghana Commercial Bank. In northern Ghana as a whole (consisting of the Upper-West, Upper-East and Northern Regions), where about 20 per cent of the total population lives, there were in 1988 only 45 bank offices, which amounted to a mere 8.6 per cent of the country's bank offices. The ratio of inhabitants to banking outlets reaches a high of over 1:58,000 with average catchment areas of 5,300 km^2.

This very low bank density seems to be one reason why the entire North accounts for only 3.9 per cent of all formal sector credit and 2.8 per cent of all formal sector deposits. . . . The insufficient supply becomes all the more obvious if one takes into account the internal concentration of banks in the three regional capitals Tamale, Wa and Bolgatanga: almost half of the bank offices in the North are located in these cities while only one-sixth of the northern population live there. This implies that the ratio of inhabitants to bank offices in 'rural northern Ghana', outside of the regional capitals, is close to 1:100,000.

(Bentil *et al*. 1988: 66)

By contrast to the recency in the evolution of banking services in the region, a phenomenon largely of the mid-seventies and eighties, co-operative credit unions have been in existence in this area since 1955.

LOCAL INITIATIVES AND THE EMERGENCE OF THE CREDIT MOVEMENT IN THE AREA

The pioneering work which led to the establishment of credit union associations in the area was undertaken by Rev. Father John McNulty among Dagaaba peasants in Jirapa parish within the Upper-West Region. This Canadian-born priest was inspired by his experience with credit unions in Canada, where they emerged in poor rural areas among the working class and craftsmen with the object of providing a collective response to their poverty through mutual self-help (Chitsike 1988: 34; Credit Unions Association [CUA] 1971: 1). He pursued the idea among the very poor parishioners in Jirapa. After several months of intensive discussions in a basic membership education programme, the first parish (associational) type of common bond credit union in Ghana was founded in September 1955 in Jirapa (CUA 1971: 1; Van Den Dries 1970: 34).

Thereafter, the credit union movement was actively promoted by the entire Catholic Church hierarchy in the Wa diocese, leading to the establishment of a credit union in every parish town. From Wa diocese the idea spread to Navrongo and Tamale diocese in Northern Ghana. The preliminary work in the parish towns was delegated to the parish priests. All but five of the twenty-two major credit unions within the Upper-West Chapter, as of 1985, were based at parish stations. For the five which lay outside mission stations, the missionaries still played an important animating role in their establishment.

As many as seven credit unions were already in operation within the North-West before the Ghana Co-operative Credit Unions Association Ltd (CUA) was formally registered in March 1968. CUA serves as an umbrella organization for all primary societies affiliated to it. CUA relates to the primary societies through the regional offices of these societies, which are known as chapters.

The role of the National Association is mainly supervisory. It aids existing societies in training officers and members; helps in the establishment of new societies; provides financial aid to member societies; represents the member societies in dealings with the government; and represents the member societies at the international level. The various chapters have the authority to act on behalf of the National Association at the regional level although this is exercised in consultation with the National Association (CUA 1978: 15).

Table 4.1 Regional distribution of credit unions affiliated to CUA 1984/5

Chapter	Credit union	Membership	Savings (C)	Loans (C)
Ashanti	29	4,913	13,984,465.00	12,531,594.00
Brong Ahafo	19	4,587	28,762,433.96	26,205,041.63
Central	15	1,748	2,776,064.03	2,527,265.63
Eastern	9	1,155	1,074,005.00	992,625.00
Greater Accra	47	9,491	21,305,677.00	17,566,489.00
Northern	7	2,010	1,809,355.00	1,624,479.00
Western	13	5,710	11,958,352.00	8,994,535.00
Volta	NA	NA	NA	NA
Upper-East	7	3,256	2,823,071.38	1,714,836.27
Upper-West	22	17,382	62,231,741.54	43,019,718.00
Totals	168	50,252	146,725,164.91	115,176,583.53

Source: CUA Office, Accra

As of the 1984/5 fiscal year, CUA embraced a total of 168 primary societies with a total membership of 50,252 persons. The regional breakdown is shown in Table 4.1. The total savings of all credit unions stood at 146,725,164.91 cedis while total loans were 115,176,583.53 cedis.

The Upper-West Chapter accounted for 13.1 per cent of the number of primary societies, 34.6 per cent of total membership, 42.4 per cent of total savings and 37.3 per cent of the total loans. The position of the Upper-West Chapter in the Co-operative Credit Union

Movement is a noteworthy achievement given the level of poverty in the region. These savings and loans have been mobilized entirely from the efforts of the rural poor.

The district by district distribution pattern of the major credit unions in the Upper-West Chapter is shown in Table 4.2. There is a concentration of societies, membership, savings and loans within the Upper-West Region from where the credit union idea diffused to the Northern Region. The two most populous districts of Wa and Lawra with the highest number of parish stations also have the highest number of societies.

Table 4.2   Distribution pattern of major credit unions Upper-West Chapter 1985[1]

	No. of societies	Membership	Savings (C)	Loans (C)
Districts Upper-West Region				
Wa	8	4,302	16,189,997.81	8,950,490.00
Lawra	8	6,597	15,199,401.16	11,019,016.80
Tumu	3	2,696	14,474,738.32	9,243,553.70
Districts Northern Region				
Damongo	1	2,029	5,379,635.99	4,290,472.10
Bole[1]	2	2,418	11,454,481.64	9,917,863.70
Totals	22	18,042	62,698,254.92	43,421,396.30

Source: Upper-West Chapter head office, Wa

1 Data for 1984 were used for Fielmun and Babile (Upper West Region) while data for 1986 were used for Bole (Northern Region) owing to deficiencies in data.

In terms of per capita savings, Tumu District tops the list followed by Bole and Wa Districts, as shown in Table 4.3. The per capita savings levels for the credit unions in Damongo and Lawra Districts fall below the mean per capita savings level for the Chapter which stands at 3,477.05 cedis. In terms of loans, the average loans per head are highest in Bole followed by Tumu. All other districts had below average performance, taking the mean value for the Chapter as a whole. The reason for lower savings rates and per capita loans in Lawra, Damongo and Wa Districts may be that the levels of loan delinquencies are higher for the oldest credit unions in these districts. This fact might have served to dampen the enthusiasm of most rank and file members in credit union activities.

Table 4.3 Per capita savings and loans by district, Upper-West Chapter

	Savings per head (C)	Loans per head (C)
Upper-West Region		
Districts		
Wa	3,763.36	2,080.54
Lawra	2,303.98	1,670.30
Tumu	5,368.96	3,428.61
Northern Region		
Districts		
Damongo	2,664.50	2,125.00
Bole	4,737.17	4,101.68
Mean for Chapter	3,477.05	2,408.02

Source: Upper-West Chapter head office, Wa

In nominal cedi terms, the capital base of the credit unions in North-Western Ghana has shown a positive trend. For example, savings grew from 16.4 million cedis in 1979/80 to just over 66 million cedis in 1986. Loan portfolios also grew from 7.5 million cedis to almost 46 million cedis in the same period (Table 4.4a). In real terms the capital base of the Upper-West Chapter showed a rather drastic shrinkage from 16.4 million cedis as savings in 1979/80 to a mere 5.8 million cedis in 1986.

This had a corresponding effect on loan disbursements (Table 4.4b). The 1981/2 financial year showed a slightly better picture because of the rigid enforcements of price controls in the first year of the December 31 Revolution when the popular classes had a strong influence on policy. This situation stands in sharp contrast to the sharp decline from 1982/3 to 1986 as a result of the very drastic devaluation of the cedi from under three cedis to one dollar (US) to over ninety cedis to one dollar (US) by 1986.

The progressive decline in the value of the cedi to the US dollar is reflected in the dollar equivalents presented in Table 4.4c. There are a number of reasons for the shrinkage in capital base of credit unions in the study area. First, the high import content of manufactured items and the limited price support for food crops which were produced in the area worsened the terms of trade of rural producers and thereby made it more difficult for peasants to sustain and increase their savings. The second major reason was the decline in

Table 4.4a   Upper-West Chapter: trends in savings and loans

Year	Savings (C)	Loans (C)
1979/80	16,441,235.37	7,575,367.26
1980/1	25,892,709.46	14,553,861.82
1981/2	36,144,337.00	19,035,228.00
1982/3	49,734,545.00	23,693,918.00
1983/4	63,617,224.06	46,766,326.49
1985	62,698,254.92	43,421,396.80
1986	66,156,635.00	45,897,307.00

Table 4.4b   Deflated values (using 1979/80 = 100)

1979/80	16,441,235.37	7,575,367.26
1980/1	11,952,315.00	6,721,547.80
1981/2	13,649,384.00	7,188,377.80
1982/3	8,428,443.70	4,013,680.40
1983/4	7,724,479.60	5,678,423.40
1985	6,896,939.60	4,776,444.70
1986	5,842,272.50	4,053,177.40

Table 4.4c   US dollar equivalent (US$)

1979/80	5,978,033.00	2,754,403.40
1980/1	9,414,588.90	5,291,783.80
1981/2	13,142,080.00	6,921,208.90
1982/3	2,446,939.60	1,165,248.70
1983/4	1,768,558.80	1,300,103.80
1985	1,153,647.80	798,953.68
1986	740,954.31	514,049.83

Source: Calculated from data collected from Upper-West Chapter head office, Wa, using Quarterly Digest of Statistics, March 1988, Accra, Ghana

confidence that peasants had in the credit unions as seen in the declining number of active members for some of these credit unions. For example, the total active membership dropped from 25,830 in 1983/4 to 16,290 in 1986.

Perhaps the most important issue relating to the development of credit unions is their role in the stimulation of endogenous rural development initiatives by these neglected rural poor, to which we now turn.

THE ROLE OF THE CREDIT UNIONS IN RURAL DEVELOPMENT

In order to address this important question, an analysis is presented below of the sectoral and spatial structure of credit distribution and issues related to class, gender and access to credit. For this purpose five credit unions, namely Wa, Hamile Town, Jirapa, Kaleo and Ko Credit Unions, were selected for closer study.[2]

In terms of effective utilization of savings as credit, most credit unions were found wanting. Out of the five selected credit unions, only the Wa and Hamile Town Credit Unions were effectively utilizing their deposits as loans to their members. Well over 75 per cent of the savings of the Wa Credit Union and all of the savings of the Hamile Town Credit Union were out in loans. By contrast, under 50 per cent of savings of the remaining three Credit Unions were given out as loans. The main reason for the difference lies in the contrasting world outlooks between the commercially minded Islamic Wala ethnic group, which dominated the Wa and Hamile Town Credit Unions, and the agrarian peasant-oriented focus of the other three unions that were located in the Dagaba area. Generally, poor peasants were found to be more averse to risk relative to the more opulent merchant class. Because of the low returns on investment in agriculture, given the vagaries of the weather, peasants living close to the margin of survival are more cautious of getting into debt. By contrast, given the highly inflationary situation which Ghana has gone through, speculative and rentier activities in the commercial sector have tended to confer windfall gains on rural merchants. More recently, the pattern has changed for the other credit unions. They now have higher percentages of their resources being utilized as loans because of the demonstrated effect over time of the advantages to be derived from access to credit and the penetration of the rural unions by urban-based elements.

The sectoral distribution of loans for all of the credit unions surveyed showed a high concentration of loan portfolios in the productive spheres, namely trade and small-scale industries, agriculture, housing, health and education. For example, over 90 per cent of the loans in Wa, Hamile Town and Jirapa were investment loans. For Kaleo and Ko Credit Unions the proportion of investment loans were 75 and 60 per cent respectively.

Within the productive sector, trade and small-scale industries together accounted for between 50 to 80 per cent of the total loans, with agriculture and housing each having a range of 3–16 per cent of the total loans from year to year. Health and education-related borrowing had the lowest percentages, generally ranging from below 1 to 5 per cent.

Only a small fraction of the loans were for consumptive purposes and these were distributed among the following categories: family use, purchase of private transport, repayment of loans elsewhere, purchase of foodstuffs and the settlement of bride price or dowry. The purchase of cars, motor bicycles and bicycles was the most outstanding single expenditure item in this category because of the high capital outlay required per unit item.

The per capita concentration of loans was much higher when borrowing was for investment rather than for consumption. The very poor members of the unions made the heaviest demands on loans for the purpose of consumption but, while they accounted for between 50 to 98 per cent of the number of beneficiaries, their relative share of total loans was generally under 30 per cent from year to year. One could conclude with some justification that the credit unions, by supporting investments in the productive base of the regional economy, have been vital instruments in rural development.

Going beyond mere sectoral aggregates, one notices: (a) a skewed distribution of credit in favour of the urban and rural centres where the credit unions are located as against their wider catchment areas; (b) a further polarization of capital in the hands of the local dominant classes; (c) a distribution of credit in favour of male members of the respective credit unions.

Taking the issue of rural/urban contrast, the privileged position of credit union centres, as opposed to the settlements in their respective catchment areas, is very stark. For example, in the case of Wa Co-operative Credit Union, all but 2 per cent of the loans went to Wa based members alone. The picture was slightly better for the Jirapa Credit Union where almost 21 per cent of total loans went to about 32 per cent of the total beneficiaries in the outlying villages. A further improvement in the position of satellite villages as opposed to the credit union station town was observed in the case of Kaleo Credit Union. About 45 per cent of the loans went to 43 per cent of the total beneficiaries in the surrounding villages. Ko Co-operative Credit Union was the only exception to the rule. Over 60 per cent of the loans was utilized by 82 per cent of beneficiaries in outlying villages.

The overall bias in favour of urban or 'rurban' centres can be explained largely by the sectoral concentration of loan portfolios in such urban activities as commerce and industry, and the distance decay effect in the spatial pattern in the relative use of an innovation by people living within the spheres of contact of the origin of an innovation. The special case of Ko Credit Union can be explained in terms of the overall rural character of the centre, apart from the existence of a Catholic mission station and the credit union. In addition, within its catchment area lay such large centres of very profitable commercial maize farming as Lambussie, where an emerging stratum of capitalist farmers has been making heavy demands for credit to finance fertilizer purchase and tractor hire services. The Lambussie area is well known in the region for 'large-scale' maize production and recently the Lambussie *Kuoro* (chief) won the 'best maize farmer of the year' award in Ghana.

In terms of the relationship between class and access to credit, loans were heavily concentrated in the hands of a few beneficiaries in the upper stratum of regional society in the cases of Wa, Hamile and Jirapa Credit Unions. In the case of Jirapa, about 60 per cent of the total loans was utilized by a mere 2 per cent of the total number of beneficiaries. For Wa, between 68 to 80 per cent of total loans was absorbed by between 6 to 10 per cent of loan beneficiaries. The situation was almost the same for Hamile where about 41 per cent of the loans was concentrated in this stratum, involving only 6 per cent of the total number of loan beneficiaries. The opposite, however, was true for Kaleo and Ko Credit Unions where nearly all of the loans went to the lower and a few middle strata of rural society. In the case of these two credit unions, per capita loans were also much smaller.

Even for these rural-based credit unions, such as Kaleo, Ko and many of the others, the current pattern is one in which the few better-placed urban and rural bourgeois elements, consisting of merchants, contractors, civil servants, chiefs and *kulaks*, i.e. farmer-traders, have seized control of the management of the unions and as a result have begun to appropriate a larger share of total loans to themselves. This tendency became ascendant in the seventies when the Catholic Church, partly in response to pressures from the local dominant classes, decided to withdraw priests from the position of treasurer in the various co-operative credit unions. The newly appointed treasurers were very often barely literate in accounting knowledge. These clerks soon became pawns in the hands of local notables and were found approving and granting loans to members of the various committees, which was against the rules.[3]

Available evidence suggests that most of the credit unions have lost huge sums of money through misappropriations. In one case alone, a treasurer misappropriated five million cedis from union funds at Damongo; at both Wa and Bole, over four million cedis each of union funds could not be accounted for.

In almost all of the credit unions, those who owed huge sums of money and who have defaulted with repayments are the chiefs, local contractors, merchants, middle-level servants and members of the various boards of management. Most of these loans were granted as 'trust' loans, without any collateral, by the loan committees. In the case of Lawra Credit Union, the repayment of loans due from two such individuals alone would be enough to put the credit union back on a very sound financial footing.

Poor peasants, rural women and workers are rarely mentioned, if at all, among the list of loan defaulters. Poor people very often seek only genuine loans, which they do pay back as they generally do not want to be disgraced. By contrast, the well-connected in society know that they can escape exposure by members of the loan committee. Rank and file members often do not understand the financial mess that their unions may be facing until such time as they cannot withdraw money from their savings because of an acute liquidity crisis.

The members of the various management boards have also undertaken poor investment projects, using union funds, with the result that such investments have served as drains for credit union resources. One good example is a hotel complex under construction at Wa, using resources from the local credit union. Others include a co-operative store and a petrol dump owned by the Jirapa Credit Union, a tractor hire service run by the Kaleo Credit Union and a fuel dump operated by the Busie Credit Union. The Tumu Credit Union has used union funds for the construction of bore-holes. All of the investments which were aimed at expanding the capital base of the respective unions are operating at a loss.

Owing to the out-migration of adult males, women outnumber men: in 1970 the ratio of males to females was 89.2:100 and this increased only slightly during the 1984 population census to 90.3:100. The discrepancy was greatest among those aged 15 to 40, the years of greatest migration. This situation is not reflected in membership lists of the various credit unions, as women often account for under 30–40 per cent of total membership. While a number of women may 'put away' their small savings in the credit unions to avoid the predatory attitudes of their husbands, most women have not benefited from the credit facilities in their respective unions. Most men in the region do not like their spouses to have financial autonomy as, in their view, it erodes their position of control. Women's chances are better when they place group savings in the credit union. The *pito* brewers' association in Ko is among a few women's groups which have utilized the loan facility of the Ko Credit Union. There are hardly any women representatives on the committees of the various credit unions operating in the area.

Although the parent association (CUA) was established in 1968, specific provision of credit for women began only in 1984 when CUA helped market women in the port area of Tema to establish the first market women's credit union in Ghana. The success of this union has led to the establishment of four more market women's credit unions, two in neighbourhoods within Tema and two more recently at Kaneshie and Nima, both in Accra. The latter two were established with the encouragement of the Canadian Co-operative Association (CCA). There are prospects for the establishment of two more women's credit unions in Accra, in Makola Market and 31st December Market (Ghana 1988: 1–2).

Similar credit initiatives are being introduced in North-West Ghana under a 'Credit for Women Project' which is being initiated by CCA in partnership with CUA. This pilot project is targeted towards three zones, namely the Upper-West Chapter in the North, the Greater Accra/Coast Area in the South and Brong-Ahafo Region in Central Ghana. The Upper-West area has been included because 'it is the birthplace of the African credit union movement, and it is currently the source of 50 per cent of the savings which have been deposited in CUA's Central Fund' (Ghana 1988: 2). This move is designed to facilitate access to credit for women credit union members for productive activities such as trading, agricultural production, food processing and marketing. It is also intended to provide the selected credit unions with the capacity for continuing management of savings and credit programmes for women and to increase women's credit union membership.

CRISIS AND PEASANT PROTESTS WITHIN THE CREDIT UNIONS

Under the emerging chaotic and undemocratic political system, many chiefs, traders, retired civil servants and public servants, who secured 'trust' loans without the usual collateral or guarantors, have refused to pay back the money, with the result that a number of these primary societies now face serious problems with loan repayment. Individuals among these groups have taken huge loans of between 300,000 to one million cedis for well over five to ten years without having repaid a pesewa. In the case of three credit unions, Hamile Town, Nandom and Wa, the liquidity problems are so severe that individual members cannot even withdraw money from their savings accounts. This has led to the collapse of the Hamile Town Credit Union.

In Nandom the situation was so grave that peasants in a fit of anger and protest literally drove off the treasurer and some other committee members. As a result, the parent credit union is facing *de facto* liquidation; peasant confidence has been completely destroyed. From the mess has emerged a farmers' co-operative credit union, the Kuob-Lantaa Credit Union. The peasants have refused to admit any literates into the new union, since the literate 'élite' were blamed for the frauds in the Nandom Credit Union.

A recent audit task force has revealed that the current liquidity crisis within most of the credit unions has been aggravated by plain theft of resources by treasurers and some members of the various committees in the chapter. About 20 million cedis have been lost in IOUs and other such fictitious loans to members of the various committees. Since the inception of these unions, no dividends have been paid to members because of staff incompetence.

Other problems relate to bad management and the overall lack of democracy within the unions. For example, in theory the supreme authority of each society is vested in the members, who exercise their power through voting at annual and special general meetings, but in practice the involvement of the general membership in the operations of these credit unions has been virtually absent. Meetings are either not held at all or are held but no elections take place, the rank and file membership merely endorsing the nominations of candidates that are made by influential members of the unions.

The checks and balances that were carefully negotiated between the various committees, namely, the committee of management, the loan committee and the supervisory committee, never worked either because of collusion among the various bodies or because of ignorance, particularly on the part of members of the supervisory committee which should oversee the operations of the management and loan committees (CUA 1988: 12–17). Most of the treasurers were selected on trust and lacked any accounting background. This was one of the main reasons why dividends could not be calculated and paid to members even though some profits accrued, largely in the form of interest on savings with the commercial banks. Auditing of accounts, if undertaken, was carried out by incompetent personnel who could not detect frauds.[4]

Summarizing, one could argue that, quite apart from managerial problems, it has been the progressive alliance between the local dominant classes and elements of the credit union bureaucracies, who themselves are aspiring petty bourgeois, that has served to exclude the poorer groups from access to credit union funds. Through the agency of the credit unions, capital is in the process of penetrating and partially dissolving the pre-capitalist economic forms in the region in its quest for a general commodification of the regional economy. By assisting the 'large-scale' commercial farmers to purchase improved seeds, fertilizers and tractors, the credit unions are encouraging the development of capitalist production relations in agriculture. Part of the loans to traders is used for the speculative purchase and hoarding of local foodstuffs at harvest time. This is often sold back to poor, distressed peasants who might have sold cheaply at harvest time only to find themselves having to buy back grain at double the harvest time price during the 'hunger season'. This tends to have a magnified negative redistributive effect on incomes. At the same time, by offering credit to traders, there is an expansion of trade in imported manufactured goods thereby helping to draw peasants more and more into the market economy and away from the production of use value in the crafts industry. A skewed income distribution also tends to expand the market for imports of luxury goods into the region, such as cars, motorcycles and imported foods. In the end, both the relative and absolute poverty levels of the poor peasantry, a large proportion of whom are women, have been deepened, giving rise to a heightened awareness of class conflict among the poor peasants.

CONCLUSION: STRATEGIES OF REVIVAL AND PEASANT CONTROL

In order to reorient the credit unions towards their initial goal of supporting the developmental initiatives of the poor majority, a number of considerations are being implemented by the management of CUA for all credit unions in Ghana. These include: (a) measures to re-capitalize the credit unions; (b) intensification of the educational training programmes for members and their mass involvement;

(c) the recovery of loans from defaulters together with improved staffing with competent personnel.

As has been indicated in the analysis, the high rate of inflation in Ghana, the devaluation by over 8,000 per cent of the cedi between 1983 and 1988, and the high levels of fraud and default in loan repayment for some primary societies, have all served to erode the capital base of the credit unions in the area. The women's credit initiative, which is being funded largely by CCA, is in the process of establishing a revolving fund of Can.$ 268,110 which is targeted on women in the three pilot zones of Greater Accra/Coast Area, Brong Ahafo Region and the Upper-West Chapter. About 30 credit unions from the three pilot areas will benefit; in addition, at least ten new, women-centred credit unions will be established in the pilot areas. All of the individual loans in the beneficiary credit unions will be for purposes of production only, and will have a maximum maturity of three years. A minimum of 10 per cent counterpart contribution will be required of all participating credit unions. The total funding involved, which includes provision for training, personnel, transport and management support, is Can.$ 429,300 with Can.$ 68,360 of this amount being CUA's contribution to the project (CCA and CUA 1988: 6–8).

As a component of the project, there is to be a greater involvement of women in the management of the various credit unions. For example, CUA will have to add at least two women to its Central Fund Committee, which will select the 40 credit unions to receive the group loans. The 40 selected credit unions will add at least two women each to their credit committees to decide which of their members will be given individual or collective loans (CCA and CUA 1988: 4).

At the end of the three-year project period the following results are expected:

(a) A total of 1,100 credit union members from the three sample areas will have received 1,400 new loans for productive purposes;
(b) A total of 960 women trainees from participating credit unions will have received general leadership and/or general business training;
(c) A minimum of ten new, women-centred credit unions will have been established as part of the 40 credit unions participating in the project;
(d) The membership of women in the 40 participating credit unions will have doubled from the beginning to the end of the project period;

(e) A minimum of 10 per cent growth in the assets of the CUA's Women's Revolving Credit Fund will have been achieved each year (CCA and CUA 1988: 7).

About 33 per cent of the total beneficiaries will come from the credit unions in the Upper-West Chapter, which will help to strengthen access to credit by women who are among the poorest members of regional society. At the same time it will improve democratic participation in the activities of the various unions, an important animation programme for women in the area.

Another initiative by the CUA is also under way with the Food and Agriculture Organization (FAO). If it is approved for support, the FAO is to assist in educating farmers in the various credit unions on marketing and credit management. Links will also be established with the Farmers' Services Centres, Ministry of Agriculture and Barclays Bank in order to pre-finance input delivery to farmers who belong to specific credit unions.[5]

The World Bank, which has been impressed by the better performance of the credit unions concerning savings mobilization in remote rural areas and by the lower loan delinquency averages compared to the banks, has expressed interest in financing small-scale industries in particular and the rural sector in general through the credit unions.[6] Repayment rates of between 92 and 100 per cent can be found in well functioning unions (Bentil *et al.* 1988: 115).

While the average loan delinquency level is below 30 per cent in the credit unions, it is estimated that the national average for banks in Ghana is above 60 per cent (Hubbert 1988: 15). Although four separate World Bank teams have recommended the channelling of financial credits to the credit unions, the disbursement of funds continue to be targeted only to rural banks.

An audit task force has been reviewing the operations of the credit unions while auditing and updating their records. A recovery of delinquent loans is part of the programme of reorganizing and recapitalizing the credit unions. In the Upper-West Chapter specifically, the Catholic Diocese of Wa has commissioned a study of the Upper-West Chapter credit unions as a basis for reorganizing and reactivating the poorly run ones (Wa Diocese 1987). If all of these schemes by CUA and the Church get off the ground, the credit unions will be sufficiently recapitalized to play a more useful role in the development of their catchment areas.

In the sphere of mobilization, the education committees of the respective credit unions in the region, which are either non-existent or dormant, need to be activated to educate the rank and file members in their general responsibilities and rights. In this regard it is necessary to enlarge the membership of the various committees to reflect the social base of the credit unions. While other financial institutions are not interested in why people save, why they want to build up capital or where the money comes from, the credit unions must be interested in people's objectives for saving, including a study of their investment plans as a way of contributing to each individual or group realizing its objectives (de Stemper 1988: 10). More credit should be directed towards group farmers, women's groups and the poorer strata of the primary societies.

Last, there is the need to strengthen the managerial capacity of the various credit unions in the Upper-West by hiring competent treasurers and officers. These are some programmes of action which will help to mitigate the harsh reality of the Structural Adjustment Programme in Ghana among the neglected rural poor in the Upper-West (Songsore 1989). Through these reforms the credit unions will be better placed to become focal points in the struggle for fundamental changes in the relations of production by poor peasants and their class allies, the urban working class, at some future date.

NOTES

1 Data from a recent survey of financial institutions in the Upper-West Region.
2 Except where otherwise stated, much of this section derives from Jacob Songsore (1982: 6–26) and further discussions with Catholic priests in the Upper-West and with the general manager of CUA, Accra.
3 Wa Diocese 1987: 1–2. Discussions with CUA general manager, priests and other people concerned with credit union work in Upper-West Region.
4 Discussions with general manager, CUA, Accra, 1988; see also CUA 1987b: 8–9.
5 Discussion with CUA general manager, 1988; see also CUA 1987a: 5.
6 Discussion with CUA general manager, 1988.

5

SURVIVAL IN RURAL AFRICA
The salt co-operatives in Ada District, Ghana

Takyiwaa Manuh

INTRODUCTION

The poverty of theories on African development has been laid bare as the crises engulfing African economies have deepened. From east to west, north and south, hopes for African 'breakthroughs' have shattered as the ravages of the world capitalist market destroyed hopes of controlling national economies, of formulating rational planning, either by state or market forces, Sender and Smith (1986) notwithstanding. Even the Côte d'Ivoire, the sometime success story of Africa, is now struggling to stay on its feet.

Documentation on the nature and extent of the crises has come from many sources, and diligent perusal of annual World Bank reports from the early 1980s, for instance, provides eloquent testimony of the manifestations of the crises, even as there are disagreements about their origins (WB 1984b, 1985, 1986). Similarly, UNICEF's reports on the state of children in Africa present the crises from the viewpoint of the most vulnerable groups – women and children in urban poor and rural communities (UNICEF 1984, 1985). More recently, Onimode (1989b) and others have presented overviews and case studies of the crises and the programmes of adjustment which have been put in place in several countries as an answer to the crises. Overall, the crises and the accompanying structural adjustment programmes (SAPs) have led to a massive deterioration in the quality of life for poor urban and rural dwellers, further deepening the rural/urban divides which exist in almost all African countries. However, the various strategies for rural development have, like national development programmes, largely failed to alleviate the deprivation that rural communities experience and most rural people have been left to their own devices. Brown (1986) and others have

described the framework, strategies and policies for rural development which have been formulated in Ghana, with little success, by
successive regimes. What is remarkable about the contributions in
the volume edited by Brown is that, with the exception of one paper
(Boateng 1986) which describes governmental and voluntary participation in rural development, most contributions focus almost solely
on theoretical issues, institutions and resources, and planning for
rural development but say little about what rural people do themselves about their situation. Yet, it is clear that, for most rural
dwellers, survival and the little development that has come their way
have largely resulted from their own efforts.

In this chapter, we present the organization of co-operatives in the
Ada Songor lagoon area (see Figure 5.1) as an instrument of struggle
and a coping strategy by rural people, in the face of dispossession by
powerful capitalist interests and state neglect.

THE ADA DISTRICT

Introduction

Ada is located in the Dangme East District of the Greater Accra
region of Ghana and covers an area of 546 km^2. Ada Foah is the
administrative capital where government offices and functionaries
are located, and Big Ada is the seat of the paramount stool and the
traditional capital of the Ada state. The population of Ada was 68,923
in 1984, of whom 36,564 were female and 32,359 male. Population
densities are generally high, with about 126 persons per km^2, which is
higher than the national average of 50 persons per km^2, and decrease
from the coast inland.

Climatic conditions

The climatic condition of the District is dry equatorial; Ada is located
between Accra and the Volta which is the driest part of the entire
West African coast. The vegetation is short grassland with small
clumps of bush or a few trees and the soils are coastal sandy, tropical
black clays and lateritic sandy soils. There are two rainfall maxima,
but the dry season is very marked and evaporation exceeds rainfall in
eleven months of the year. Mean temperatures are around 27°C and
mean annual rainfall is between 737 mm to 889 mm. Given these
features, the scope for agriculture is limited to short term crops or

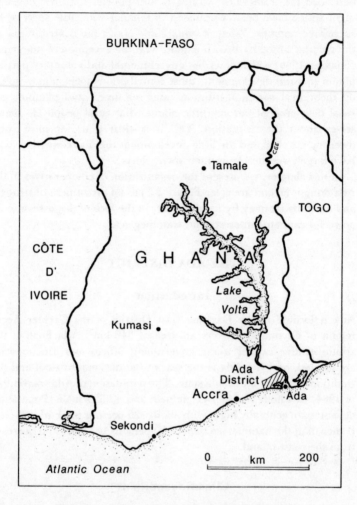

Figure 5.1 Map of Ghana showing the study area

crops which can survive on intermittent rainfall, and only about 47 per cent of the arable land is under cultivation at any one time. The main crops grown are cassava, maize, tomatoes, okras, peppers, bambara beans and groundnuts, with the latter four crops inter-planted on cassava, maize or tomato farms. Table 5.1, which shows agro-economic subzones, population, arable and grazing land in the Ada zone, adequately portrays the characteristics of the zone and

Table 5.1 Ada agro-economic subzones, population, arable, grazing and total areas

Zone description	Total area (ha)	No. of people	Arable areas (ha)	Grazing areas (ha)	Arable area per person (ha)	No. of cattle	Grazing area per cattle unit	Remarks
1 Coastal fishing villages	6,160	8,734	460	1,150	0.05	336	3.42	West Songaw probably includes grazing used by saltings subzones
2 Islands and estuary	1,130	3,111	765	—	0.25	—	—	
3 Market and urban	650	9,156	negligible	negligible	negligible	—	—	Some farming in adjacent areas
4 Fishing and farming coast, East Songaw[1]	3,860	4,604	405	3,080	0.09	1,556	1.97	Includes Midier grazing
5 (a) Angaw[1] Valley, Gorm (b) lower (c) upper	4,840	5,259	2,295	1,880	0.44	1,111	1.65	
6 Marsh grazing	6,110	3,831	495	5,220	0.13	2,299	2.27	Both Obanes included
7 Kasseh, Toje Red Sands	7,640	6,172	5,490	1,190	0.89	441	2.70	Probably some grazing on arable
8 Hwakpo-Koluedor Sege	7,100	4,989	3,110	3,350	0.62	1,152	2.91	
9 NW inland grazing	15,750	1,946	770	15,140	0.40	4,757	3.18	Probably grazing outside area
10 Saltings	1,100	2,168	negligible	510	—	1,514	0.34	Must be grazing outside area
11 Lagoons, rivers	11,170							
Totals all zones	65,510	49,970	13,790	31,520	0.29	13,178	2.31	

Source: Ada farming households survey, 1983

Note: [1] 'Songaw' is also spelled 'Songor' in some texts; similarly 'Angaw' and 'Angor' are interchangeable spellings.

consequent settlement and occupational patterns. At least 45 per cent of the total area is covered with marshes, saltings, lagoons and rivers, which effectively sets the limits of economic activity. A significant feature of the Ada District is the high ratio of livestock, arising from the fact that the zone is relatively free from the tse-tse fly. Thus, in spite of constraints of water and marginal grazing land, cattle-rearing is an important economic activity. In addition to livestock, there are large flocks of sheep, goats and many pigs.

The Ada Songor zone: demographic and socioeconomic characteristics

Of the 68,923 persons in Ada District, about 24,152 persons or 35 per cent live around the Songor lagoon in some twenty villages. Of these, 12,924 are female and 11,228 are male. Tables 5.2, 5.3 and 5.4 show the age and sex composition of the population, economic activity and regular school attendance for the zone.

Table 5.2　Age and sex composition, Ada Songor zone 1984

Sex	All ages	Below 1 year	1–4	5–9	10–14	15–24	25–44	45–64	65 and over
M	11,228	312	1,858	2,066	1,288	1,981	1,992	1,048	683
F	12,924	376	1,980	2,235	1,351	2,225	2,522	1,440	795
Total	24,152	688	3,838	4,301	2,639	4,206	4,514	2,488	1,478

Source: Ghana 1984

Table 5.3　Economic activity, Ada Songor zone 1984

Sex	Total age 15 and over	Employed		Unemployed	Homemaker	Other
		Total	Agriculture, hunting, forestry and fishing			
M	5,704	4,987	4,339	36	37	644
F	6,982	6,317	1,678	29	130	506
Total	12,686	11,304	6,017	65	167	1,150

Source: Ghana 1984

Table 5.4 Regular school attendance, Ada Songor zone 1984

Sex	Total age 6 and over	6–14 years			15–24 years			25 years and over		
		Never	Past	Present	Never	Past	Present	Never	Past	Present
M	8,577	1,584	150	1,146	757	758	466	2,546	1,170	—
F	10,114	2,158	124	850	1,473	563	189	4,194	563	—
Total	18,691	3,742	274	1,996	2,230	1,321	655	6,740	1,733	—

Source: Ghana 1984

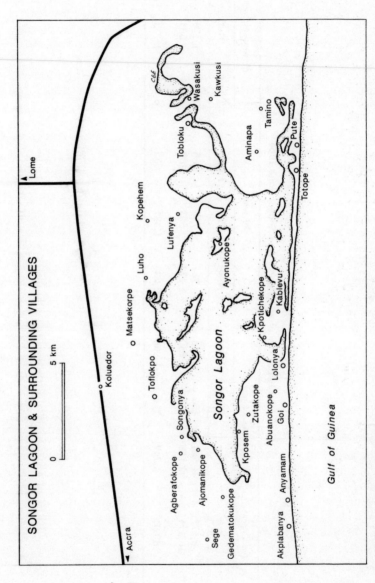

SONGOR LAGOON & SURROUNDING VILLAGES

0   5 km

▲ Accra

○ Koluedor

▲ Lome

○ Matsekorpe

○ Toflokpo

Agberafokope ○

Songonya ○

○ Ajomanikope

Songor Lagoon

Kposem ○

○ Zutakope

Sege ○

Gedematokukope ○

Abuankope ○

Goi ○

○ Lolonya

Akplabanya ○

Anyamam ○

○ Luho

Lufenya ○

Kopehem ○

○ Ayonukope

○ Kpotichekope

○ Kablevu

Cêê

○ Tobloku

Wasakusi ○

○ Kawkusi

Aminapa ○

Tamino ○

○ Pute

Totope ○

Gulf of Guinea

Figure 5.2 Map of Songor Lagoon and surrounding villages

Settlement in these villages dates back almost to the time when the Adas found themselves in their present homeland. The Songor lagoon area is inhabited by Tekperbiawe, the clan which traditionally 'owns' the lagoon, and other Ada clans. The lagoon provided fish in the rainy season and salt in the dry season and has centrality of place in the cultural and religious lives of the villagers. These villages have been linked to many places inland through the production of salt in the lagoon which has attracted traders from far and wide. In addition, trade in smoked fish was carried on with towns as far away as Mangoase, Akuse, Agormanya and other places on the Volta (see Figure 5.2). In return, the traders, mostly women, carried back cocoyams, palmnuts, bananas, oranges and other food items which were unobtainable in the Ada environment. In addition to trading, other Ada migrated with their families in the 1920s to the cocoa growing areas in Akim and around Koforidua in order to establish cocoa farms or to work as caretakers on the farms of wealthier farmers. A few also went to forest sites to log timber in order to make canoes and boats.

The main occupations of people in the Ada district are farming, fishing, livestock rearing, and trading, with the majority of employed people working in agriculture and related activities including fishing (Table 5.3). Surprisingly, salt mining is not listed under economic activities although it formed and remains an important occupation for villages around the lagoon. In part this may be attributable to the time of year when the census was conducted, when the inhabitants may have been engaged in other activities; it could also reflect the ecological changes which occurred with the construction of the Volta Lake and which blocked the entry of sea water into the lagoon and made salt winning a precarious occupation; lately, it might also have reflected the conflicts over access to the lagoon after private salt winning companies came in, depriving local people of access.

At present there is only one industrial establishment in the Songor zone, Vacuum Salts Limited, which employs about two hundred people on a casual basis. In addition, the salt co-operatives also employ a number of young people as waybill clerks and secretaries. The remaining salaried employment is by the central government and district council which employ teachers. The overwhelming majority of the inhabitants are self-employed farmers, fishermen and saltwinners, individuals often combining two or three occupations at the same time. In addition to these usual occupations, a few men are tailors, carpenters, shoemakers, fitters, distillers of *akpeteshie* (the local

109

gin), cassava mill operators, weavers of mats for fencing compounds and ceilings, makers of straw mattresses, kraal owners, and carvers. Some women trade in both local produce and industrial products, sell cooked food on an itinerant basis or in the small local markets, are chopbar keepers in villages along the main road, are processors of cassava into dough or *garri* for sale in Kasseh and Accra markets, are sellers of drink, or are weavers of baskets for transporting crabs, a speciality in the Ada district. Most of these occupations are usually combined with farming, fishing or salt winning. Since these occupations yield variable returns, migration is widely adopted as an alternative.

Infrastructural facilities and community amenities

Water supply

In all of the villages, the existence and level of basic amenities is very low.[1] Water is a major problem for most villages around the lagoon as the level of salinity in the soils is very high. To compound the problem, there are few streams flowing into the lagoon, thus making most villages dependent on dug-out ponds which dry out quickly or which are easily contaminated by cattle and other animals. All communities report that they buy water during at least three months of the year for domestic and other uses. Water thus constitutes an important item of household expenditure.

Health facilities

In addition to problems with water, which also pose a threat to health, the lack of adequate health facilities is a major constraint facing the inhabitants of the Songor zone. The nearest health facility is more than 30 km from most villages. There is dependence on traditional birth attendants (*tbas*), herbalists and rudimentary community clinics.

Energy sources

There is no electricity in any of the villages surveyed, nor indeed in any of the villages around the lagoon. Vacuum Salts Limited has a generator but electricity does not extend to the village near where it operates. The major sources of energy are fuelwood and charcoal for cooking and domestic uses, and kerosene for lighting.

Educational facilities and enrolments

Since the early 1950s, primary schools have existed in nearly all of the villages and middle schools have been added to some of these.

Table 5.5. presents a summary of the basic services and facilities that are in place in six selected villages.

THE ADA SALT TRADE

Extensive documentation of salt production and marketing indicates that Ada salt has been produced and traded for over two centuries. Isert, a French traveller in the 1780s, gave a detailed account of the salt industry at Ada in 1784. He wrote: 'To each house at Ada were attached huts each of which could store at least fifty tons of pure salt' (Isert 1793, quoted in Dickson 1969). According to Isert, the Adas looked down on agricultural activities, preferring instead to engage in salt mining and fishing. They sold the products to the people inland and bought food from their neighbours. Dickson (1969) also states that 'the salt works at Ada [were] probably rather large and important, since there were few profitable alternative occupations other than fishing in the Volta and neighbouring creeks'. Clearly, then, the reasons why Ada people engaged in fishing and salt mining lay more in their natural environment than in preferences. As the above description of climatic conditions showed, this is an area of lagoons and marshes, backed by inland grassland plains and supporting little agriculture.

Ada salt was traded throughout the country, and Bosman (1705, quoted in Dickson 1969) claimed that the salt trade rivalled that of gold in profitability, with a handful of salt worth one or two slaves. Together with fish, the salt was transported to Kete Krachi where it was sold to trading caravans from northern Ghana (Dickson 1969: 180). Ashanti was also an important market for salt from Ada, carried there along the great east–west routeway that linked Kumasi with the oracle Odente at Krachi.

The Ada salt trade continued and survived throughout the colonial period in spite of the competition from imported salt and the imposition of taxes. Ada salt was preferred to other varieties of salt on the market and competed favourably, despite higher transportation costs. Sutton (1981) gives an informative account of the trade, its marketing network around the country, and the virtual monopoly of the Ada traders. While salt collection was regarded as a task mainly for

Table 5.5 Amenities in six sample villages in Ada Songor Zone 1988

Main activity by village	Population 1984	Access by road (main, feeder)	Vehicle based in village	School (primary, middle, junior secondary)	Health facility[1]	Water supply[2]	Tractor in village	Mill	50% of buildings with metal roofs	50% of buildings with block walls	Artisan (carpenter, fitter, tailor)	Production co-operative
Marine fishing												
Lolonya	2,284	F	yes	P,M	CC,HB,TBA	R,P,T	no	yes	yes	yes	C,T	yes
Anyamam	2,739	F	yes	P,M,JS$	CC,HB,TBA	PB,W,T	no	yes	no	yes	C	yes
Agriculture/ livestock/saltwinning												
Koluedor	2,572	M	yes	P,M,JS$	CC	P,R,T	yes	yes	no	no	C,F,T	yes
Matsekorpe	825	F	yes	P,M	HB,TBA	P,W,T	no	yes	—	—	—	—
Adjomanikorpe	700	F	no	—	HB,TBA	P,R	yes	yes	no	no	no	no
Lufenya	430	F	no	P	HB,TBA	W	no	yes	no	no	no	yes

Source: Survey data, field observations

1 CC = Community clinic; HB = Herbalist; H = Hospital; TBA = Traditional British attendant.
2 R = River; P = Pond; T = Tanker; PB = Pipe borne water; W = Well.

women and children, control of the trade was in few hands, as investment was required in the form of canoes and hired paddlers. Later, more successful traders such as the Ocansey and Dumas families bought steam launches for navigation up the Volta. The salt trade formed an important, albeit not the sole, source of accumulation for such traders (Moxon 1984).

In the past, the process of collecting salt from the lagoon demonstrated community management of a natural resource. The Volta enters the sea at Ada Foah and, as Sutton (1981: 50) recounts, before the construction of the dam, the Volta and channels leading to the lagoon flooded their banks around September–October. As flooding retreated, brackish water was left in the lagoon and, as the sand bar was narrow and fragile, sea water could also enter at high tide. In any case, the water would be saline, as the Volta was tidal for some kilometres upriver. If the lagoon dried out before the rains came, about 2.5 cm of salt was left. When this occurred, the Libi Wono and Woyo, priest and priestess of the Tekperbiawe clan, would be informed. At the direction of the Ada *mantse*, the paramount chief, they would place sticks and watchmen on the paths leading to the lagoon and would cause an announcement to be made, placing a ban on salt winning in the lagoon. This measure ensured that no one collected the salt before it was all exuded, and gave equal access to all. It further maintained the process of collection firmly in the hands of the priests and the chiefs, who levied tolls on salt won. By January, the process would have occurred and the lagoon would be left to dry out for about three months. When the priests and elders determined that the salt had sufficiently matured, the ban would be lifted and people could then go out to win salt. Tolls were collected on the salt and were shared between the paramountcy in Big Ada and the priests.[2]

From all accounts, salt did not form in the lagoon every year. However, given the size of the lagoon, which was 13 km long and 8 km at its widest, even occasional production from the entire lagoon would result in storage of large amounts from year to year.[3]

From about 1960 the lagoon began to fail. This was in part caused by the creation of the Volta Lake and the construction of Akosombo dam which had deleterious effects on the lagoon, leading it to dry out more frequently. At the same time, the entry of sea water into the lagoon was impeded both by silting at the mouth of the Volta and by the control of flooding that was required for the dam to generate hydro power (Anim 1959; Sutton 1981: 50).

113

In the 1970s, two private companies, Vacuum Salts Limited (VSL) (owned and managed in principal by the Appenteng family) and Star Chemicals Limited, were given leases by local chiefs which were subsequently confirmed by the government, to win salt from virtually the whole expanse of the lagoon. A part of the lagoon was reserved for the Ada Traditional Council. This area was considered sacred by the inhabitants as the abode of the goddess of the lagoon. However, from the point of view of the inhabitants, the area was unsuitable for salt winning. It received hardly any sea water for the necessary concentration to be achieved for forming salt. In addition, the area was far from most of the villages.

In the early 1980s, conflicts erupted between the lagoon dwellers and the two companies over access to the lagoon. In particular Vacuum Salts Limited, had constructed dykes and embankments across the lagoon and had stationed armed guards there to prevent the villagers from exercising any rights to the lagoon. Although the actual operations of the companies were outside the lagoon (for example, Vacuum Salts possessed several pits which were used as crystallizing pans, these being fed by sea water which was pumped in through a pipe), the companies claimed monopoly rights over about two-thirds of the lagoon. Ada people were arrested on charges of stealing salt, were locked up in cells and were constantly routed by the police who acted on behalf of the companies. It was during this period that a salt co-operative, the Ada Songor Salt Miners' Co-operative Society, emerged to organize the saltwinners and to act as a focus in the struggle to regain rights over the lagoon.

In 1985, the government finally intervened in the conflict when a pregnant woman was shot and killed in a raid. A committee of enquiry was set up to investigate the grievances of the inhabitants arising from the operations of the companies, to ascertain whether these operations had resulted in any hardships to the people, and to make necessary recommendations.[4]

THE SALT CO-OPERATIVES

In Ada, as in other districts in Ghana, there are four main types of co-operative societies which reflect the predominant occupations and the needs of members for inputs for production: co-operative farming societies, fishermen's co-operatives, bakers' co-operatives, and *akpeteshie* distillers' societies. These societies usually flourish in periods of scarcity and reflect the constraints that farmers, fishermen, bakers and distillers, for example, face in their productive activities. Their main aims are to secure inputs, in the form of fertilizer, cutlasses, outboard motors, nets, flour, sugar and loans, for members. They are registered with the Department of Co-operatives, operated under the latter's regulations, and depend on it for access to inputs.

However, in the Songor zone, the salt co-operatives are new and different. Although they have utilized the co-operative form and regulations, they are different in conception and function. These co-operatives arose from the struggle of Ada people, predominantly the inhabitants of the lagoon area, to regain sovereignty over the lagoon. Originally, one co-operative society, the Ada Songor Salt Miners' Co-operative Society, emerged in 1984. Currently there are some ten societies, most of whom operate provisionally under licence from the Department of Co-operatives, but by far the best organized is the original society. Its attempt to emerge as an engine of development in the zone will now be considered in detail.

The Ada Songor Salt Miners' Co-operative Society

In 1983, the People's Defence Committees (PDCs), which had been formed in communities and workplaces following the events of 31 December 1981 in Ghana, took over the operation of Vacuum Salts Limited (VSL) while Star Chemicals Limited was confiscated by the state. By 1984, the situation had reversed; the former management of VSL under the Appenteng family resumed their position and subsequently barred access to the lagoon by the inhabitants. The potential for improving the livelihood of the inhabitants that had been demonstrated during the administration by the PDCs led to the mooting of the co-operative idea by some local activists and their allies as a means of combating VSL. In January 1984, contacts were made with leading personalities in the villages around the lagoon. The co-operative concept was explained to them, resulting in member groups being formed in some six villages to win salt. By October–November 1985, when the government announcement was made which allowed anybody to win salt in the lagoon, membership of the Co-operative comprised 3,200 people or about one-sixth of the total population of the Songor zone.

At first, the general need of the population for money led to very low prices for salt in Ada. A 75 kg bag of salt cost only C 40 in Ada but sold for C 600 in Accra, only 80 km away. Attempts were therefore made to increase the price through levying a toll on each bag. This money was to go into a development fund to benefit the area. A customs duty was also exacted, which was paid directly to the government, and the Co-operative issued waybills and customs receipts to buyers. In time, personnel from the Customs, Excise and Preventive Services were stationed at buying points around the lagoon to collect duties. By 1986, which the Co-operative reckons as its best year of operations, its assets totalled some C 15 million (US$ 100,000 at 1986 rates). Most of this revenue was put into treasury bills for a year. This choice of investment was due to a large extent to the role played by a director and leading figure in the co-operative movement who was an accountant. Such involvement by a professional raises possibilities about the roles that African professionals can play in sustaining and facilitating the development of rural communities.

Ironically, the financial success of the Co-operative was to evoke demands from sections which had played little or no part in its struggles. The first of these was the District Administration in Ada Foah which was faced with a financial crisis following the government's cuts in subventions to district councils as part of the Structural Adjustment Programme on which it had embarked. District councils had to meet 50 per cent of the salaries of their staff from the beginning of 1987, and so the Ada District Council passed a by-law in 1986 which authorized it to collect a toll of C 60 on every bag of salt sold. As has been shown above, the villages around the Songor benefited little from the operations of most governmental departments in the District capital. Thus, initially, the Co-operative refused to pay. Later, the Co-operative unwillingly paid one million cedis on account because of the harassment of its customers by militiamen acting at the instigation of the District Council. In spite of the payment, the District Council itself issued waybills to certain villages around the lagoon to collect tolls on its behalf, on the understanding that they would share the money in the proportion of two to one. The formation of other co-operative societies around this time further undermined the Ada Songor Co-operative's stand against the Council, since these other societies acceded to the Council's demand to pay tolls. The Ada Songor Co-operative finally agreed that the District Administration could station personnel on their premises to collect the tolls.

The Ada Traditional Council was the other body which began to make demands on the Co-operative after the District Council succeeded in collecting its tolls. The Council insisted that the Co-operative should pay a toll of C 50 per bag for the development of Ada and the upkeep of the chiefs, as royalties for the use of the lagoon. Although the Co-operative refused to accede to their request, it decided as a gesture of goodwill to make contributions to the annual Asafutufiam festival in 1986. As part of this contribution, two Kumasi Ventilated Improved Pit (KVIP) latrines were started in 1986 at Big Ada and completed in 1987, at a cost of C 1.1 million paid for by the Co-operative, with the help of National Service personnel and communal labour by the townspeople. However, these contributions did not appease the Traditional Council and since 1988 it has used the Ada Traditional Area Fund (ATAF) to market salt and to collect tolls on its behalf.

The Ada Songor Co-operative Society has a management committee made up of 24 members, seven of whom are elected, while the rest are observers. This committee meets monthly to discuss developmental and other objectives of the society, as well as to deal with any problems which may arise. There is also an 18-member secretaries' committee, which is made up of the secretaries and their assistants from constituent villages, which meets on an *ad hoc* basis. Finally, there is the general meeting which is supposed to meet fortnightly, but which in reality rarely meets. This creates the paradox of a body which, although well organized at the top, is almost non-existent at the grass-roots level. The explanation given, that most women who are the majority of members are interested only in winning salt and have no time to participate in meetings, can only be partial. More fundamentally, it reflects not only the failure, so far, to translate the co-operative idea into reality for individual co-operators in the peculiar circumstances of the Songor area but also the internal contradictions which have emerged from within the ranks of the co-operative society itself. Thus, while co-operatives of the old type were concerned primarily with enhancing the benefits to their co-operators, the Ada Songor Society has set itself broader development aims and objectives which can only be realized over the medium to long term, and not necessarily at the individual level. The realization of these objectives, however, demands the strong participation of members, and the inability to involve rank and file members in its activities further delays progress. Part of the inability to integrate members stems from the male-dominated structures. So far, only one woman stands out alongside the men even though salt winning itself is an occupation dominated by women. In addition, the scheduled

117

meeting times do not coincide with women's time-use patterns and do not, as a result, enhance their participation.

Problems with conceptualizing and realizing the potentials of this new type of co-operative have been compounded by allegations of financial irregularities, embezzlement and non-accountability within the Ada Songor Co-operative Society. This has occasioned inter-necine strife and thrown rank and file members into confusion and suspicion, leading to the mushrooming of societies in virtually every village. At the same time the Co-operative's funds have been blocked. The Department of Co-operatives, through its District officer, appears not to have been of much help and to have no faith in the capabilities of the salt co-operatives. Beyond recommending that the co-operatives be grouped in zones and that a curb be put on the number of co-operatives which can be formed in the area, the Depart-ment has not been in a position to enforce its own regulations concerning the formation of societies. Regular audits have not been performed as required by law, nor have annual general meetings been held, which are to be attended by officers of the Department. More recently, the Dangme East District Secretary has called for a re-organization of the co-operative system around the lagoon, and for the Department to perform its statutory duties.[5]

Despite these difficult problems, the Ada Songor Co-operative Society can claim successes in the initiation of development projects in some of its constituent villages, albeit at a slower rate than projected. In general, development projects are determined by felt needs of the people, and the provision of water has been identified as the major problem. However, given the high salinity and the sandy soils, it has not been possible to dig many wells, as the well casings easily settle. World Vision International, an international voluntary development organization, has been approached to help with the construction of bore holes in villages around the lagoon, and it is expected that this will begin shortly.

The construction of classroom blocks for existing schools is seen as another priority to enable the schools to function well, and the Society has provided roofing sheets, timber, cement and other materials for constructions in Matsekorpe, Koluedor and Anyamam, while towns-people have provided communal labour. There are also plans to put up KVIP latrines around the banks of the lagoon for salt winners. Some pits have been dug but are not completed. Finally, in order to secure vehicular access to sites along the lagoon, the Society has engaged in road maintenance and in the construction of culverts.

Erosion is heavy around the lagoon owing to the high salinity, and the articulated trucks which ply around the lagoon, also put a heavy strain on the existing roads. However, following the formation of the other co-operatives, the Society is unwilling to bear all of the costs of road maintenance, and so it has proposed a co-operative union to discuss services and cost-sharing for the use of the lagoon as well as general management of the lagoon.

The Society has also constructed offices for its own use and is now building warehouses to store salt for an export drive to Nigeria and neighbouring ECOWAS countries. It has formed a marketing company, African Salt Limited, to be in charge of this, and the company has already started operations.

THE STRUGGLE OVER THE SONGOR LAGOON AND POWER RELATIONS IN ADA

In the 21 July 1989 issue of the *Daily Graphic*, it was reported that the president of the Ada Songor Salt Miners' Co-operative Society, who is also an Assemblyman, had been dismissed from the Assembly following a letter he had written headed 'Attitude of PNDC District Secretary for Dangme East and Officials of the Department of Co-operatives towards the Ada Songor Salt Miners' Co-operative Society Limited'. While the legality of the purported dismissal of the people's nominees by an assembly remains to be tested, the above incident illustrates the mounting contradictions between, on the one hand, the co-operative societies as new engines for development around the Songor zone and, on the other hand, the interests of the District Administration and chiefs who have long considered the lagoon as a source of revenue for development in the District and traditional capitals of Ada. This section discusses the interplay of forces, both local and external, who are dominant actors in the continuing struggle to define the rights of access to, and the use of, proceeds from the Songor lagoon.

The State and the Songor lagoon conflict

The role of the State in the conflict over access to and control of the lagoon has depended on which forces have been able to mobilize the apparatus of the State to their side. Initially, in 1982, when the PNDC came to power professing to speak for the masses, the then People's Defence Committees (PDCs) were sufficiently emboldened

119

to take control of the salt companies' premises and operations and to administer them. However, this control was ended in 1984, when a government announcement was made which restored the management to Vacuum Salts Limited and Star Chemicals to their owners. Effectively, however, only VSL resumed production while Star Chemicals shut down. The leadership of the PDCs was castigated and accused of maladministration and embezzlement. Following the ascendancy of the attacks by government on external and internal exploiters, and the adoption of an economic recovery-cum-structural adjustment package, the scope for private capital expanded somewhat in order to secure the confidence of much needed investors. In this climate of *laissez-faire*, it was important for Appenteng, VSL and other investors to be assured of governmental non-interference with their entrepreneurship. Indeed, the State intervened on their behalf, as is shown in accounts of workers' struggles in the PNDC period, from the turnover of the Ghana Textile Printing Company to the case of the Assene metal workers (see, in particular, Graham 1989).

During the period 1984–5, the full power of VSL and the Appentengs was unleashed on the people of Ada, who had their salt seized in markets at Kasseh and Dawa and in raids on their homes (Memorandum presented to the Amissah Committee of Enquiry by Ada Songor Salt Miners' Co-operative Society, 1985). Salt was seized at barriers erected by the police, acting on the instigation of VSL, on the pretext that it was stolen. The fledgling Ada Songor Salt Miners' Co-operative Society had its offices raided several times by customs officials and had its books taken away. Watchmen and other employees of VSL brutalized Songor residents who were found near the lagoon and confiscated implements used in winning salt. Some members of the Appenteng family were accused by residents of personally administering cruel and inhuman punishment such as forcing men and women to drink concentrated salt water from the lagoon.

Finally, police detachments were used to intimidate, assault and batter Songor residents, resulting in the notorious raid on 17 May 1985, at the lagoon site near Bonikorpe village, in which a pregnant woman was shot dead. In addition, other residents sustained injuries. At this point, the State intervened, this time on the side of the Songor residents. An inquiry was ordered into the grievances of the residents, and access to the lagoon was granted to all, saving actual trespass on the premises of VSL.

But the State's intervention was only partial, and not definitive, as since the submission of the Amissah Committee of Enquiry Report of Ghana (Ghana 1986), the findings have not been published. Further, the State has revalidated the leases to VSL, without any publicity, and has thus undermined the findings of the Amissah Committee on the operations of the two salt companies. In addition, technical teams have been appointed which have sought to denigrate the operations of local saltwinners, thereby setting the stage for banning local people from winning salt. For example, the report of the Technical Team appointed by the Ministry of Lands and Minerals Resources in July 1988 proposes that the lagoon should be exploited by companies using more 'scientific' methods to win salt.[6] However, as evidence before the Amissah Committee of Enquiry clearly showed, there was little, if any, difference between the technologies used by local winners of salt and by the two private companies: for instance, Vacuum Salt Limited, contrary to its name, was not using methods of vacuum extraction in its operations but, like the local winners, private companies depended on the over-abundant sunshine to collect salt from its evaporation pans.

In these circumstances, it is apparent that factions within the State apparatus are mobilized in support of VSL and its claims, and are sometimes able to further VSL's aims. However, while the contest may be unequal, residents of Ada Songor are not without their champions. From time to time, seemingly unexpected announcements are made to their benefit. Thus, in October 1989, the management contract was terminated between the government, as leading shareholder, and VSL, as represented by Appentang & Co., and the Appenteng family were banned from the area as a result of continuing acts of harassment (*Daily Graphic*, 30 October 1989). But, given the equivocal position of the State *vis-à-vis* Songor, the State cannot be relied upon to protect the interests of the residents. This was apparent recently in Ada when a demonstration was organized to protest against the extortionist acts and bias of the District Secretary against the salt co-operatives. It resulted in the arrest of a militant priest who was released only after intervention from Accra. However, populist measures do have their value and, from time to time, the State can be relied upon to come to the defence of Songor residents, under strong pressure from lobbyists.

The salt co-operatives and local authority holders

The salt co-operatives have suffered not only from the power of the State, which often mobilized support of VSL, but also from the power of local authority holders. As adverted to above, control over salt production in Ada has been a matter for dispute among priests, local notables and the Ada *mantse*, and, according to Sutton, conflict raged during the colonial period over claims to particular shares.

The role that the chiefs and local notables played in the acquisition of the Songor lagoon by the two companies appears not to have been salubrious; a charitable view could be that they were impressed with the prospects for developing the zone that were held out by the companies. Even then, they were soon disenchanted and refused to accept the royalties proffered, on the grounds that they were too paltry. With the emergence of the salt co-operatives and their generation of funds, the chiefs, through the Ada Traditional Council, have insisted on what they see as their due and have demanded allowances for their upkeep and contributions to other causes. While the cooperatives have sometimes been obliged to keep the peace, there has also been conflict. The chiefs used the Ada Traditional Area Fund (ATAF) to market salt on their own behalf, as distinct from the interests of the Songor residents.

The chiefs have allied themselves with the District Administration to challenge the hegemony of the Ada Songor Salt Miners' Cooperative Society, and have not been averse to acts of intimidation. The role of the presiding member of the District Assembly, who is also chair of ATAF, is being investigated. He is currently suspended from duty.

For the chiefs and notables in Ada, it would appear that the lagoon exists as a source of accumulation for development in Big Ada and in Ada Foah and its environs, but not around the lagoon itself. Thus, it would appear that in part the present conflict centres on whether this historical function will be maintained or whether the lagoon will serve more directly the interests of the Songor zone as a whole. The existence of other co-operative societies with less well-defined aims and objectives has been used to force the Ada Songor Salt Miners' Co-operative Society into line. Its own organizational weaknesses militate against action to confront the chiefs and other local interests successfully.

The needs of the District Administration for a local source of revenue to meet the costs of decentralization, in the absence of strong central government support, have placed it on a collision course with the salt co-operatives. The failure to engage in dialogue and a consensual approach to resolving matters results in the issuing of edicts and commandist postures which serve only to polarize positions and to create antagonism. So far, few benefits of development, as pursued by current or past district administrations, have accrued to the residents of the zone and notice has been served that, instead of waiting, the people are mobilizing themselves to deliver what they need.

CONCLUSION

Will the co-operative societies survive all of these onslaughts? It is clear that they face serious internal organizational problems which have to be resolved effectively if this new co-operative concept is to have any salience for the membership. Effective participation by the rank and file, especially women, accountability and probity of officers, and a more cohesive organization need to be built up in order for the co-operative to carry out its developmental tasks with greater success. In a situation of plurality of co-operatives, co-ordination needs to be enhanced among the several co-operatives and common tasks shared to avoid unnecessary and unhealthy duplication and rivalry. Such understanding may well pave the way for the merging of nonviable co-operative societies to ensure viability. The unity that derives from this co-operation could then be used to confront the interests and forces which are marshalled against the co-operatives and could lead to the development of the potential of the area to provide a better life for the inhabitants.

NOTES

1 The lack of social amenities and infrastructure in Ghana, particularly in rural areas, has been noted in the Ghana Living Standards Survey, which was funded by the World Bank, and some measures are proposed to deal with this issue. See WB 1988a.
2 Sutton (1981: 50) alludes to conflict between the priests and the Ada *mantse* over the division of tolls, and to contradictory accounts about the actual division of tolls.
3 Sutton quotes Hutter, who had been the agent of a company at Ada since 1896, on the volume of salt production from the lagoon (Sutton 1981: 51). Nowadays, customs and excise proceeds show the volume of salt that is traded in a season, although it is still difficult to tie these figures to production in any particular year.

4 This was the Amissah Committee of Enquiry set up by Executive Instrument No. 12 of 1985.
5 This is with reference to the address by the District Secretary at the first ordinary sitting of the Dangme East District Assembly where he called for the reorganization of the co-operative system being operated by salt-winners in the Songor lagoon, the financial backbone of the district. This was to correct the situation whereby a society comprised only the executive, and he asked for the submission of full membership lists. See *Daily Graphic*, 27 April 1989.
6 This is a reference to the 'Report (interim) of the Technical Team for Songor lagoon, presented to the PNDC Secretary for Lands and Mineral Resources, Accra, Ghana, July 1988'.

6

LOCAL COPING STRATEGIES IN MACHAKOS DISTRICT, KENYA

Peter O. Ondiege

INTRODUCTION

Africa's economic structure defines the essential features of her central problem of underdevelopment. The basic bottle-necks that arise from structures of production, consumption, technology, employment and socio-political organization include problems of mass poverty, food shortage, low productivity, weak productive base and backward technology. The fundamental problem of Africa is that of

> a vicious interaction between excruciating poverty and abysmally low levels of productivity in an environment charac-terised by serious deficiencies in basic and social infrastructure, especially the physical capital, research capabilities, techno-logical know-how and human resources development that are indispensable to an integrated and dynamic economy.
>
> (OAU 1986: 4)

The crisis is exacerbated by factors that affect economic performance. These include inflation, export earnings instability, balance of payments deficits and rising debt burdens which are the direct result of the lack of structural transformation, the rather unfavourable physical and socio-political environment of African economies and their excessive outward orientation and dependence.

Africa's economic structure is characterized by a predominantly exchange economy, narrow production base, neglected informal sector, environmental degradation, urban bias of public policies and openness, and excessive dependence on external factor inputs.

Most African countries depend on agriculture for employment, income, foreign exchange earnings and government revenue, and yet agriculture is still characterized by traditional production techniques and a generally low level of productivity. Most efforts to transform this sector have been concentrated on the export subsector, ignoring the food and raw materials subsectors. While women play an important role in agricultural production, especially the food subsector, their role as producers and agents of change in the much-needed rural transformation has been severely constrained by their meagre share in the means of production (land, capital, credit, technology, etc.) and by their marginalization in production relations. Thus, owing to the weak agricultural base, the industrial sector has also remained structurally weak and narrow and with insufficient internal linkages.

African economies have also tended to ignore the informal sector which is currently estimated to account for 20 per cent of total output and over 20 per cent of the total labour force. This sector plays an important role in production, distribution, finance and employment creation in the African economies and needs, therefore, to be given serious consideration to help to reform Africa's economic structure.

Natural and human factors have continued to erode the natural resource base which is needed to sustain Africa's development. These factors include over-use and misuse of the soil, poor conservation policies, overgrazing, deforestation, drought, salination, river system pollution and water-logging. The cumulative effect of these factors has led to reduced carrying capacity of the land at present levels of technology, reduced productivity, social dislocation and accentuation of absolute poverty in the rural area.

The past development strategies that stressed industrialization and urbanization have failed to sustain growth of these economies. Objectives of providing dynamic forces for the structural transformation of their economies and promotion of equitable growth have not been successful.

In sub-Saharan Africa, the rural population seems to be the most affected as even minimum basic needs have not been adequately provided. Thus, in these economies, national development should be essentially concerned with rural development. This entails promotion of agricultural and non-rural farm activities, and increasing economic and social welfare levels of the population through employment creation, rising income and poverty reduction. The ultimate aim of rural development should be to improve the earning capacitities of farmers and off-farm workers and also to provide services and living conditions which are better adapted to their needs. To achieve this, it is imperative to provide an enabling environment that will allow the participation of local rural people in conceptualizing, implementing and managing their development programmes.

In 1948 Kenya's population was 5.25 million which increased to 8.6 million by 1962. By 1979 the population had reached 15 million and it is expected to rise to 24 million in 1989. The majority of these people are living in rural areas with only 15 per cent living in urban areas. The population is estimated to be 35 million persons by the turn of the century, with an estimated urban population accounting for only 25 per cent of the total (Kenya 1989b). Essentially, then, the rural areas will still have a majority of the population.

Table 6.1 Growth rates of real gross domestic product (GDP) 1964–87[1]

Year	Agriculture	Manufacture	Government services	Other	Total GDP
1964–71[2]	4.2	8.2	9.8	6.9	6.5
1972	7.6	7.3	12.8	3.6	6.8
1973	4.4	14.4	6.3	1.0	4.1
1974	− 0.2	5.9	6.8	4.0	3.1
1975	4.6	4.0	8.5	− 0.01	3.1
1976	3.7	14.0	5.1	2.0	4.2
1977	9.5	16.0	5.1	6.1	8.2
1978	8.9	12.5	6.4	8.4	7.9
1979	− 0.3	7.6	7.1	7.7	5.0
1980	0.9	5.2	5.6	5.2	3.9
1981	6.1	3.6	5.3	6.9	6.0
1982	11.2	2.2	3.8	1.4	4.8
1983	1.6	4.5	4.2	1.5	2.3
1984	− 3.9	4.3	2.9	2.7	0.8
1985	3.7	4.5	4.2	1.5	4.8
1986	4.9	5.8	6.3	5.4	5.5
1987	3.8	5.7	5.7	4.9	4.8

Source: Kenya 1989b: 5

1 1964 prices for 1964–71 and 1982 prices for 1972 onward.
2 There were extensive revisions in major series affecting GDP calculations in 1972. Any linkages between 1964 and 1982 based series should therefore be interpreted with these revisions in mind.

The rapid population increases have resulted in high pressures on land such that land subdivisions and inequities are growing. In addition, there are rising employment problems and pressures on other resources. Growth rates of agricultural production in Kenya, which averaged 4.2 per cent per annum between 1964 and 1971, increased to 7.6 per cent in 1972 and then started to decrease until 1976 (see Table 6.1). Farm production is also affected by weather changes; for instance, in the 1983/4 drought, agriculture fell by 3.9 per cent.

Kenya has also been faced by unemployment programmes since independence. The historical urban unemployment has been 16.2 per cent (Kenya 1989b) and is not expected to change in the coming few years. In rural areas, the employment problem mainly has been underemployment and not open unemployment. There is therefore a need to encourage those activities that enhance employment in rural farm and non-farm activities as well as those of urban informal activities where underemployment problems exist. These, if enhanced, may increase rural incomes, reduce poverty levels and thereby increase gainful employment.

In Kenya's rural sector, increasing population pressure on limited arable land has resulted in a large number of small landholdings which are insufficient to support household production (Table 6.2). These co-exist with large, underutilized holdings. This has led to poor agricultural growth rates. There has been continued pressure on other resources and unemployment and underemployment problems have been increasing since independence. These have resulted in low income earnings and poverty of the rural households, which seem to have increased with population growth rates.

There is need to emphasize the urgency for implementing human-centred development programmes which imply that mass poverty must be alleviated in order to increase people's welfare. Development has to be engineered and sustained by the people themselves through their full and active participation, and should not be undertaken on their behalf. It may be argued that the major cause of failure of development models of a macro nature in Africa is the inadequate attention and recognition given to local micro initiatives in rural areas. In rural economies local people have developed different strategies to cope with development problems. These include ecological, socioeconomic and political strategies.

A number of African countries have of late realized the significance of local initiatives in coping with development problems in rural areas and are now trying to incorporate them in their planning strategies. Kenya, like many other countries in the Third World, has been pursuing policies with mixed results which are aimed at promoting rural development. Past policies have not been very successful and so, in 1983, the Kenya government shifted the responsibility of planning

Table 6.2 Proportion of households/holders by size of holding

District	No holding	Up to 2.0	2.1 – 4.0	4.1 – 6.0	6.1 – 8.0	8.1 – 10.0	10.1 – 15.0	15.1 – 20.0	Over 20.0	Total
					Holding size in acres					
Kilifi	—	36	22	12	10	4	14	1	1	100
Kwale	—	30	26	15	12	4	5	3	5	100
Taita-Taveta	—	52	16	10	5	3	6	4	4	100
Machakos	—	30	15	15	6	10	8	5	11	100
Kitui	6	17	20	13	7	7	15	3	12	100
Meru	2	47	32	9	4	3	2	1	0	100
Embu	1	42	25	12	6	9	4	0	1	100
Nyeri	1	62	18	11	5	0	1	0	2	100
Murang'a	—	65	21	5	3	2	3	0	1	100
Kirinyaga	—	51	24	15	8	2	0	0	0	100
Kiambu	8	68	16	4	2	2	0	0	0	100
Nyandarua	2	27	21	10	9	3	7	5	16	100
Nakuru	22*	25	23	10	4	7	3	4	2	100
Nandi	6	25	16	10	9	8	12	6	8	100
Kericho	—	25	22	10	6	7	14	7	9	100
Uasin-Gishu	3	28	11	10	5	10	11	10	12	100
Trans-Nzoia	7	45	19	12	5	5	2	1	4	100
S. Nyanza	2	47	14	8	4	5	4	6	10	100
Kisii	1	58	25	9	1	2	1	3	—	100
Kisumu	—	59	22	9	5	4	1	—	—	100
Siaya	—	42	28	14	6	3	3	—	4	100
Kakamega	—	59	23	5	5	3	2	2	1	100
Bungoma	2	27	32	14	6	5	9	2	3	100
Busia	—	23	29	20	9	5	6	6	2	100
Mean	5	40	21	10	5	5	5	4	5	100

Source: Kenya 1989a: 28

*Mainly workers on large-scale farms and estates but including farming members of land-buying companies whose farms had not been sub-divided at the time of the survey.

and implementing rural development from the headquarters of relevant ministries to the districts. This strategy of 'District Focus for Rural Development' (DFRD) reflects recognition of the importance of involving the local people in the development. Its purpose is to broaden the base of rural development and to encourage local initiatives in order to improve problem identification, resource mobilization, and project design and implementation. A further objective is to encourage and increase communications between the local community and government officers working in the district.

The success of this DFRD strategy is debatable. One issue concerns how much economic, administrative and financial powers should be given to the districts in order to enhance the successful implementation of the strategy. On paper, the DFRD strategy provides a policy environment that can accommodate local initiatives and strategies for rural development. It is too early to assess the full implications of the strategy.

This chapter examines local coping strategies of Kenya's Machakos District. The main objective is to identify and analyse the coping strategies that people have adopted in view of development problems in this district.

BACKGROUND OF THE DISTRICT

Machakos covers an area of 14,250 km^2 and is situated in a transition area between the rather dry, south-eastern parts of Kenya and the wetter, high central part that borders Nairobi City. It is predominantly a semi-arid area with a bimodal rainfall pattern. Rainfall in the area is inadequate and unpredictable, a factor which, when combined with shallow soils, steep slopes and unstable surface soil structures, makes water and soil conservation for farming a delicate issue. On the basis of agro-ecological potential, Machakos is divided into low, middle and high potential zones which constitute respectively 56.2, 38.4 and 5.4 per cent of the agricultual land.

The District has been known as a 'Problem District' in Kenya since the colonial period. The earliest problem was overstocking, later it was soil erosion, and then overpopulation. The population is now estimated to be 1.5 million and is concentrated on the more fertile parts of the central hill masses.

The consequence of the variable rainfall pattern is that, in four years out of ten, there is a major drought (Were and Akong'a 1981). For instance, in 1930–1 there was major flooding and between 1940 and 1944 there was famine. In the period 1951–2 heavy rains destroyed crops, resulting in famine and in 1965–6 there was another drought. In 1980, the food crop failed, although people claim that they had money to purchase food. In 1983–4 a major drought in the country led to food shortages which affected Machakos District severely. Generally poor soils and the arid climate contribute to low agricultural potential, which makes smallholding rainfed farming difficult in more than 75 per cent of the District. Soil and water conservation are necessary in order to have adequate farm production.

During the colonial period, there was a breakdown of the farming system cycle of land use and regeneration. This took the form of squatter settlement in unpopulated areas of Ukambani and common grazing land of the Yatta plateau. A shorter fallow system resulted (Meyers 1981). There was increased bush clearing, felling of timber and the subsequent demarcation of new holdings. At the same time, the population was expanding resulting in a growing fragmentation of landholdings. By the 1930s, land tenure was increasingly under individual control. The combination of these factors led to intensified soil erosion by the mid-1930s.

After the Second World War, poor agricultural production and famine relief became the order of the day and the government had to send in relief assistance to the District. The Akamba who occupy the District faced resettlement as well as soil and water conservation programmes which were instituted by the government. Within five years, soil erosion control was so successful that farm production increased, enabling the District to be a net maize exporter. However, this situation was unsustainable in the long term as the programmes had failed to integrate cropping and livestock components of the farming system.

In the late 1970s, Arid and Semi-arid Lands (ASAL) became a major concern of the independent government. The Machakos Integrated Development Programme (MIDP) was started in 1978 by the government, in co-operation with the European Economic Community (EEC). The focus of the MIDP is on resettling people in the former scheduled areas (of large-scale holdings), on crop and animal husbandry promotion, on the expansion of community services, and on income and employment promotion. MIDP is also concerned with soil and water conservation as well as providing essential infrastructure and water supplies.

Population, employment and incomes

The Akamba ethnic group occupies the District. The group is concentrated mostly on the more fertile parts of the central hill masses. Population densities decrease in a south-easterly direction where rainfall is less in both quantity and reliability (Tables 6.3 and 6.4). Machakos District had a higher population growth rate of 4.2 per cent than the national average of 3.9 per cent according to the 1979 population census. The current population is estimated to be 1.5 million people with the high potential zones still having high population growth rates and hence increased densities.

Table 6.3 Population projections by Division 1979–90

Division	Residential area (ha)	Population (000's)			
		1979	1988	1989	1990
Central[1]	727	82.2	121.1	125.7	130.5
Kathiani	1,069	74.4	109.5	113.7	118.1
Kilome	1,323	152.4	224.4	232.9	241.9
Mbooni	535	92.6	136.4	141.5	147.0
Makueni	2,005	125.9	185.4	192.4	199.9
Kibwezi	3,400	98.9	145.7	151.2	157.1
Yatta	2,459	137.2	202.1	209.7	217.8
Kangondo	598	133.0	195.8	203.2	211.1
Mwala[2]	1,332	125.5	184.8	191.7	199.2
Total	13,448	1,022.1	1,505.2	1,562.0	1,622.6

Source: Kenya, Machakos (1989a)

1 These divisions did not exist at the time of the 1979 census.
2 Residential area: where people can settle, excluding water bodies.

Table 6.4 Population densities per Division (persons/km^2)

Division	1979	1988	1989	1990
Central	113	166	173	180
Kathiani	69	102	106	110
Kilome	115	170	176	182
Mbooni	180	255	265	275
Makwueni	63	93	96	100
Kibwezi	29	43	44	46
Yatta	56	82	85	89
Kangundo	222	328	339	353
Mwala	94	139	144	149
District density	76	112	116	121

Source: Kenya, Machakos (1989a)

Most of the labour force in the District is self-employed. Total wage employment, according to the Central Bureau of Statistics (CBS), was 41,293 (1986) with the two towns of Machakos and Athi River in the District accounting for 33 per cent of all wage employment. Labour in wage employment was 7 per cent in 1980, only rising to 8.7 per cent in 1988, and is expected to fall to 8.4 per cent by 1993 (Ondiege 1989).

This implies that over 90 per cent of the District's labour force is employed in farm and non-farm activities in the rural areas, predominantly in smallholding farming and pastoralism.

The main source of income in the District, as in many others in the country, is agriculture, which accounted for 50 per cent of the total according to the 1982/3 CBS Rural Household Budget Survey (Kenya 1988). Off-farm enterprises, salaries and wages, and other sources respectively accounted for 17, 24, and 9 per cent. The average monthly income, including cash and in-kind, was Kshs 864 in 1981/2. The nominal rural per capita income had decreased to Kshs 628 per month, according to estimates from the rural incomes in the District for the year 1986, although income in the urban centres in the District averaged Kshs 1,860 per capita (Ondiege 1989). The situation is likely to foster more migration into urban centres in the District thereby leading to increased urban unemployment.

The decreased nominal rural per capita income between 1981/2 and 1986 is an indication of a growing population that outstrips the growth of agricultural and livestock production. In response, farmers in Machakos are diversifying their sources of income through improved agricultural practices, the adoption of new crops, and non-farm activities. Often such activities have been carried out by self-help and women's groups, the objective being both to increase food production and to generate income to improve living standards.

Access to social facilities such as health have been affected by the increasing population. For instance, the ratio of medical staff to population increased between 1983 and 1987. This implies that a medical officer working in 1983 had to attend to more patients in 1987 than he/she did in 1983 (Table 6.5), a situation that is likely to affect the health standards in the District. One solution to this problem is to train more medical staff and to intensify primary health care (PHC) programmes in the District, since most of the major diseases in the District (Table 6.6) could be prevented if nutrition and water facilities were improved. The ratio of hospital beds to population between 1983 and 1989 has improved, however, though few new health facilities were constructed between 1979 and 1987. Infant mortality rates have decreased from 109 deaths per 1,000 live births in 1979 to 65 in 1987 (Ondiege 1989).

Some of the health facilities and those for education have been built on a self-help basis by the local people, to supplement government efforts, in the realization of the deteriorating situation in the District.

133

Table 6.5   Medical services ratio 1983 and 1987

	1983	1987
Projected population	1,239,822	1,500,000
Bed population ratio	1:1,196	1:1,149
Medical personnel/population ratio		
Medical officers	1:29,520	1:34,091
Kenya registered nurses	1:10,891	1:18,519
Enrolled community/enrolled nurses	1:2,109	1:4,389
Clinical officers	1:23,393	1:20,270
Public health officers	1:72,931	1:136,364
Public health technician	1:14,417	1:16,304
Pharmacists	1:59,039	1:71,429

Source: Kenya, Machakos (1989a)

Table 6.6   Major disease incidence 1987

Disease	Number of patients	Percentages of total
Malaria	397,989	26.53 (28.62)[1]
Disease of respiratory system	310,008	20.66 (22.0)
Disease of skin (including ulcers)	115,577	7.70 (9.63)
Diarrhoeal diseases	74,973	4.99 (5.00)
Internal worms	57,848	3.85 (3.96)
Ear infections	39,707	2.64 (—)
Eye infections	31,836	2.12 (2.20)
Rheumatism, joint pains, etc.	25,624	2.00 (1.9)
Pneumonia	26,966	1.79 (—)
Accidents (including fractures, burns, etc.)	26,811	1.78 (2.05)

Source: Kenya, Machakos (1989a)

1 Numbers in (   ) are 1983 percentages.

COPING STRATEGIES

The District is faced with land shortage, population pressure, physical resource base deterioration and the vagaries of an erratic climate. The Akamba have adopted a number of ecological, social and economic strategies to cope with this problem. Some strategies were started before the MIDP was launched and represent initiatives which were undertaken by the local people themselves.

These strategies are undertaken by individuals and, more importantly, by organized groups in the locality. In Ondiege's survey (1989), the main ecological strategies included bench terracing and tree planting, dam construction, afforestation and water conservation. The economic strategies included farming, goat, poultry and bee keeping, ranching, dairy farming and fishing. People were also involved in employment and income-generating activities such as construction of rental housing, constructing and operating shops and storehouses as well as operating book shops and restaurants. There were also savings and credit, consumers', and farmers' co-operative societies. Social strategies were geared towards home improvement and welfare in general, whereby members contributed money to help one another. Groups were also involved in the construction of education and health facilities.

There was a total of over 115 Self-help Groups (SHG) and Women's Groups (W/G) which were engaged in ecological activities. In economic activities there were 96 SHGs, 249 W/Gs, 73 Farmers' Co-operative Societies (FCS), and 19 Savings and Credit Societies (S & C) in the District. Some of the SHGs and W/Gs combined these activities with social ones. Most of the groups were formed in the last decade, although these activities were carried on previously in order to cope with the deteriorating rural situation in the District. Many of these organized groups are registered with the District authorities. Although co-operative organizations existed in the 1960s, a number of them were registered after independence.

Structure, organization and strategies of the groups

Most of the women's and self-help groups are basically organized at the village and sublocational levels. Membership is mainly female for SHGs and W/Gs have only women as their members. The analysis of several groups will serve to identify some of their main characteristics.

The Kundi ya Kuela Mbesa W/G, based in Kisuini village of Machakos municipality, started in 1983 and has a membership of 28 women. One of the members says that 'the local people thought of the idea because of their needs, so they wanted to get together to solve their problems and meet their needs. There was no support from outside.' This is a W/G whose members contribute money to a common pool from which members can borrow at a reasonable interest rate. The group is facing problems of late payments of contribution and poor records of repayment. As a result, the group has no plans for expansion, which may be construed as implying poor management.

Another group, Ngenda SHG, whose chairperson is a woman, says that the initiative originated from the local people. The group was formed in 1981 and has 16 members. The main activity is the running of a shop but members also engage in organizing weddings and in making group presents to members' families. The members are planning to purchase their own plot where they will construct shop premises. This implies that they are not currently experiencing problems. In contrast, Kanzokea W/G in Kangundo Division started in 1978 and has 50 members who carry out bench terracing, horticulture and also contribute money to the group. All of these activities were started by the local people with some advice from people outside the locality. The group is said to have declined in the 1970s, owing to poor leadership, but was revived in 1981. It is currently facing financial problems.

In Yatta Division, 31 women's groups, who share a tradition of farming and weaving sisal baskets, amalgamated to form Yatta South Women's Group Enterprise Development (YSWGED) in 1986. However, the individual groups have been in existence since 1978, each being headed by a woman. Membership of each group ranges from 18 to 74 individuals, aged between 16 and 100 years old. The majority are illiterate.

The basic aim of amalgamating in Yatta was to mobilize women for socioeconomic activities, to create awareness of their potential and to raise living standards. The main focus of YSWGED is to transport members' handicrafts and grains to Yatta Women's Centre and to collect water, thereby lessening the demands on women's labour; to advertise in order to promote handicraft sales; and to use scales to ensure accuracy when buying and selling grains. YSWGED is run by an elected committee with representatives from each of the 31 groups. Their staff is composed of a general manager, storehouse manager, handicraft workers, grain store workers and labourers and strappers (employed on casual terms), and two watchpersons, all of whom are women.

YSWGED is one of the most organized women's groups in the sample. Since unification, the group has completed construction of their multi-purpose centre. Member groups have also started small projects such as group farming, building lodgings, shops and meeting places, rabbit rearing and purchasing plots. They participate in community projects such as building nursery and primary schools, soil conservation and road construction projects. YSWGED has been appointed as an agent of the National Cereals and Produce Board which purchases and markets cereals that are produced in the entire country. The group has had technical support from the Danish Volunteer Service and financial support from the Kenya Rural Enterprise Programme. In both cases the YSWGED approached the two donors.

In Mwala Division, Ndubi W/G started in 1969 with basket making and later bought a *posho* (maize) mill. In 1970 they embarked on buying plots on which they constructed rental housing to generate income for members. The group's activities were initiated by the local women themselves without outside help. They sell their baskets locally and lack organized marketing channels.

In Central Division, Ndulu W/G, which started operation in 1978, focused mainly on rental housing construction and basket making, with the objective of income generation. In Mbembani W/G, the activities are mainly poultry keeping and rental housing construction, so that members are able to diversify the sources of their incomes. These activities are meant to capture the growing market in Machakos Municipal Council where these groups are located. Wittu wa Mwene W/G, which started in 1952, has as its main activities home improvement, money contributions to improve members' welfare, and coffee growing.

It can be seen from the comments above that leadership of these groups is by women. The groups have initiated activities that may lead to improved rural development, as activities address local needs, i.e. food production, employment and income generation. The groups are found practically in all locations of the District and membership numbers are between 10 to 100 SHG and 15 to 100 per W/G. The majority of W/Gs which were surveyed had between 20 and 60 members per group.

The SHGs have a mixed membership of women and men with the former dominating. Ondiege (1989) found that most of the activities of these SHGs were socio-ecological, although there was a tendency for each group to engage in all types of activities. For instance, the main activities of Manzaa SHGs in Kilome division were the construction of dams and water tanks to conserve water and the growing of horticultural crops, i.e. vegetable and fruit, while Kyanguli SHG was involved in tree planting to conserve soil, citrus fruit growing, and money contributions by members for welfare activities.

The women's groups are said to have contributed the lion's share in the mobilization of communities at the grass-roots level, especially in the implementation of self-help projects and income-generating projects (Kenya, Machakos 1987). Women's groups were mainly involved in soil conservation, shop keeping, trade, storehouse keeping, goats, pigs and bee keeping, basketry, maize milling, tree planting and construction of dams. There were over 249 W/Gs involved in economic activities in the District (Ondiege 1989).

Women are viewed as increasingly significant participants in socio-economic and ecological conservation activities in the District in response to the deteriorating rural situation. In addition, the formation of large numbers of these groups, which are controlled and managed by women, also indicates an attempt to gain access to and control over resources that have historically been dominated by men. Women have actively taken the leadership of these groups so as to 'create awareness of their potential, raise living standards so that they may continue to develop and help their families and the community' (Kenya, Machakos 1987). They may have found it more expedient to form groups rather than to try to do it individually, as it is easier for them to compete effectively against men as a group rather than as individuals in order to obtain access to and control of resources.

Initially, about 1945, co-operative societies were restricted to coffee co-operatives which were started by white settlers. More recently, co-operatives have been formed by handicraft makers and sand harvesters in the District. The latter, mostly men, collect sand from riverbeds and load it on trucks for sale not only in the District but also, more importantly, in Nairobi where the construction industry is growing very fast. In Mwala Division, Wamunyu Handicraft Co-operative Society was formed in 1965 with the objective of promoting wood carving and basket making and marketing. This group has had mixed results but has also helped to reactivate the old crafts industry in the District.

The significance and potential of these 'coping strategies' for enhancing rural development is discussed below.

SIGNIFICANCE AND POTENTIALS OF COPING STRATEGIES

The strategies that are adopted by these local people have been on the increase and have contributed highly to the local economy (Kenya, Machakos 1987). To turn first to ecological issues, soil and water conservation and afforestation are vital in an environment that is fragile and prone to ecological disasters; local strategies supplement government efforts in this sphere. For instance, in Kaiti area, 13,850 seedlings were planted and 991 ha of bench terracing were constructed. In Tawa, Mbooni division, 13,309 ha of land were terraced, 5,180 ha cultivated and 9,791 ha of bush cleared. In Kilome Division 46,850 ha were terraced, while in Kathiani Division 6,443 ha of land were terraced and 43 ha cultivated. All of these were done on a self-help basis.

Bench terracing helps to check soil erosion when sudden and short rains fall after long spells of dry season. This is reinforced by afforestation activities which are undertaken by women and men of the area. These activities help to improve agricultural productivity which in turn helps to maintain minimum subsistence and to improve the communities' nutrition levels and living standards.

The annual report for the District by the Ministry of Culture and Social Services (Kenya, Machakos 1987) estimates that the self-help projects which were undertaken by the local people contributed about 54 million hours of labour worth Kshs 8.5 million, cash worth Kshs 2.9 million and materials worth Kshs 2.2 million, totalling Kshs 13.6 million for the year. These organized SHGs and W/Gs have helped in mobilizing local physical and human resources when implementing their socioeconomic coping strategies. They have also led to the revival of local crafts and industries and to the emergence of new institutions and economic activities in the area.

The new economic activities in the area include the construction of rental housing, shops and *posho* mills. New institutions that have emerged are the co-operative societies and women's groups, some of which have formed development enterprises such as the YSWGED. Basket making and wood carving ventures by W/Gs, SHGs and co-operative societies have led to the revival of old crafts and industry in the District.

Akamba are famous for wood carving, which is said to have been introduced in Wamunyu area around 1920 by one of the Akamba. This activity has mobilized human resources and export sales for the region and the country thereby earning foreign exchange for the economy. The Wamunyu Handicraft Co-operative Society has contributed to increasing employment opportunities for the local people, to income earnings and export sales. The Society estimates that the handicraft industry in the area employs about 2,000 persons in the location, out of which 700 are members. In 1987 the share capital was Kshs 11,250 and the turnover, Kshs 939,963. The Society has opened a showroom at Wamunyu trading centre and, in a way, has boosted the growth of the centre's trade and other trading centres in the area.

Kenya's *cyodos* (sisal baskets) and wood carvings are famous in European and American markets; the biggest share in terms of production of *cyodos*, as well as wood carvings, comes from Machakos District. During 1987, Machakos District Handicraft Centre in Machakos town assisted the women's groups by buying products locally and selling them overseas for a total of Kshs 374,185. Most Akamba use the proceeds from handicraft sales to augment family incomes, which helps to maintain the minimum subsistence needs of the local people. They combine handicraft making and farming, concentrating on handicrafts during the off-farming seasons and thus providing employment throughout the year. With proper marketing, handicrafts could increase local people's incomes far more than is currently the case, as at present great portions of the profit are received by middlemen (Mukoko 1987).

Farming activities have improved food production in the District, raising incomes of those growing cash crops and helping to improve people's living standards (Tables 6.7–6.10). For instance, in 1987, the Co-operative Societies which are mainly found in Central and Kangundo Divisions had a total of 150,087 members with a share capital of Kshs 74.97 million. The year's turnover was Kshs 287.85 million with Coffee Co-operative Societies contributing Kshs 164.81 million. In 1988, coffee earning in the District was Kshs 290 million, showing the great contribution that the local farmers were making to the economy.

Available statistics for 1983 and 1986 show that crop production and hectarage for most crops have been increasing (as may be seen in Tables 6.7 and 6.8) in the years between 1983 and 1986. There has been a great increase in the growing of traditional food crops such as sorghum, millet and cassava, which are suitable crops for semi-arid areas. Yields per hectare have also increased as a consequence of the adoption of new farming techniques by individual and group farmers, as discussed earlier.

In Table 6.9 we note that prices increased in 1984, decreased in 1985 and from then on increased in nominal values. The price increase was a consequence of the drought of 1983/4 which affected the whole country, resulting in food shortages. However, good weather conditions in 1985/6 contributed to food price reductions as a result of bumper harvests.

Table 6.7 Crop production trends 1983–6

Major crops	Metric tonnes		% change
	1983	1986	
Food crops			
Maize	121,940	178,780	47
Pigeon peas	5,840	15,840	139
Beans	69,850	56,100	– 20
Cowpeas	10,790	20,410	89
Green grams	1,060	3,320	213
Sorghum/millet	2,210	5,440	146
Cassava	6,050	19,180	217
Horticultural crops			
Bananas	18,600	16,280	– 12
Cabbages/kales	32,290	96,150	198
Onions	430	15,560	3,519
Tomatoes	12,690	126,470	897
Citrus	8,200	55,250	574
Passion fruit	720	860	19
Pawpaw	23,800	26,400	10
Irish potatoes	—	29,760	—
Industrial crops			
Coffee	22,688	28,421	25
Cotton	4,890	2,400	51
Sunflower	130	450	246

Source: Kenya, Machakos (1989a)

Similar trends are noted in Table 6.10, owing to the same 1983/4 drought. The effect of the drought may have influenced farmers to double their efforts when weather conditions changed. They also illustrate the vulnerability of a farming system which is based on rainfed agriculture and the need to improve on irrigation schemes in such regions of the country if productivity is not to be adversely affected. If farmers manage crop production efficiently, it is estimated that the average gross margin per day for coffee growers is Kshs 50 in the central hill mass areas, while for those in lower areas, who are growing cotton with sunflower as main cash crops and maize, beans and pigeon peas as food crops, the equivalent figure is Kshs 18. Thus, coffee farmers would be wealthier than other farmers if they were to manage their farming efficiently, other things being equal.

Table 6.8 Crop area yield trends 1983–6

Major crops	% change in ha 1983–6	Yield/ha (Kshs) 1983	Yield/ha (Kshs) 1986
Maize	32	900	1,000
Pigeon peas	70	450	630
Beans	13	900	830
Cowpeas	35	450	630
Green grams	162	450	540
Sorghum/millet	120	570	640
Cassava	98	5,000	8,000
Bananas	45	6,300	9,000
Cabbages/kales	32	6,640	1,500
Onions	1,386	6,143	14,962
Tomatoes	534	18,940	29,758
Citrus	192	6,508	15,014
Passion fruits	20	14,400	14,962
Pawpaw	11	2,164	2,131
Irish potatoes	—	—	8,000
Coffee*	9.7	1.3 kg/tree	1.5 kg/tree
Cotton	60	162	200
Sisal	1	1,211	1,372

Source: Kenya, Machakos (1989a)

Note: Coffee is given in kg/tree, not a cash equivalent as coffee is determined based on an average kilogram yield per tree.

Table 6.9 Price trends for food crops 1983–8

Food crop	Unit of sale (kg)	Kshs 1983	1984	1985	1986	1987	1988
Maize	90	220	320	270	280	270	290
Pigeon peas	80	360	520	400	450	540	700
Beans	80	480	600	480	520	540	540
Cowpeas	80	380	480	400	430	450	560
Green grams	80	640	720	620	700	790	720
Millet	80	600	720	620	630	640	720
Sorghum	80	340	420	380	320	480	315

Source: Kenya, Machakos (1989a)

Table 6.10   Price trends for cash crops

Period	Coffee (millions)			Cotton (millions)	
	Production cherry (kg)	Turnover (Kshs)	Net payments to farmers (Kshs)	Production (kg)	Value (Kshs)
1983–4	22.6	148.6	78.6	4.4	20.2
1984–5	16.9	174.5	130.7	2.1	9.9
1985–6	28.4	270.9	226.2	7.5	34.1
1986–7	25.6	92.5	78.1	1.3	6.2

Source: Kenya, Machakos (1989a)

The growth of citrus fruit and vegetables increases the fruit and vegetable supply and consumption in the area and also generates income for members of these groups. This helps to diversify food production and intake in the area and to improve the health conditions of the local people. According to the District Commissioner, Machakos District Co-operative Union has been licensed to export horticultural produce. This will boost production, increase income for farmers in the District and also earn the country foreign exchange.

The production and hectarage of horticultural crops have increased in the period between 1983 and 1986 (Tables 6.7 and 6.8) and yields per hectare have risen greatly. For example, the yield per hectare of onions and citrus fruit has more than doubled (Table 6.8). This expansion is likely to increase further as the co-operative society concerned has been authorized to export the produce, reflecting their importance in the District. This will increase farmers' income and improve their living conditions. Fruit and vegetable production is also important when enhanced with irrigation schemes, especially smaller scale schemes that farmers can handle easily.

Livestock farming contributes to beef and milk production, both of which have a ready and wide market in nearby Nairobi City. New production techniques, such as cross-breeding, and access to veterinary training improve the production of cattle. Livestock production has generally been increasing in the District as seen in Tables 6.11 and 6.12. However, in 1984, owing to the effects of the 1983/4 drought, production decreased except for those animals that do not rely heavily on rainfed farming, such as dairy cattle, which can be zero-grazed, and improved poultry. Milk production, however, decreased in 1984. The most drastic decrease occurred among goats and sheep which are reared for the most part in the drier, southern parts of the District.

Table 6.11   Livestock population trend ('000s)

Type	1980	1982	1984	1986	1987
Beef cattle	395.1	435.6	314.0	322.4	338.5
Dairy cattle	10.1	16.7	23.3	30.6	31.9
Goats	437.3	551.5	235.0	217.0	227.0
Sheep	187.5	298.5	82.9	18.3	91.7
Pigs	0.1	0.2	0.4	0.2	0.2
Improved birds	25.0	28.0	34.5	47.9	53.0
Local birds	1,000.4	975.0	235.0	468.0	590.1
Rabbits	2.6	5.1	4.0	7.9	9.0

Source: Kenya, Machakos (1989a)

Table 6.12   Milk production and revenues 1980–7

Year	1980	1981	1982	1983	1984	1985	1986	1987
Total milk production (million kg)	1.1	1.1	2.1	2.2	1.7	3.9	5.4	5.7
Revenue from sales (million Kshs)	3.2	3.2	6.2	7.8	6.6	13.0	15.2	15.4

Source: Kenya, Machakos (1989a)

This trend of production shows how reliance on rainfed livestock in semi-arid areas can have disastrous effects. It shows the importance of water availability, hence the need to conserve the ecology and diversification of farming techniques and sources of income if living conditions of the local people are to be maintained and improved.

Rental housing construction by these groups helps to alleviate housing shortages in the area and also assists in increasing the quantity and skills of artisans working in the construction industry. This activity, together with the construction and running of shops, storehouses and restaurants, will create jobs and diversify the sources of income for local people.

It has been noted above that W/Gs helped in mobilizing communities at the grass-roots levels in the implementation of self-help projects, especially those for generating income. About 48 per cent of the activities of W/Gs were economic in nature and 23 per cent were ecological. Women's and self-help groups augment the economic activities that are carried out by individual men and by co-operative societies which are dominated by men in terms of leadership. These societies contribute greatly to the local economy in terms of food production, trade and other activities which generate income and employment. Those that experience managerial and organizational problems need to be supported by the authorities concerned so that achievement of their potential, like that of women's and self-help groups, can be facilitated to the maximum.

A number of groups have plans to expand and some have already expanded and diversified, such as Mbembani W/G, which has started construction of rental housing over and above poultry keeping, and Katelewa FCS, which has bought a farm with cattle on it at Namanga. The Yatta South Women's Group Enterprise Development has also expanded its socioeconomic activities, as discussed previously. The potential of these micro-level groups, in terms of contributing to the economy and rural development in Kenya, is great; an enabling environment which recognizes the micro-level activities that are initiated by the local people is essential to ensure that such initiatives are sustained.

CONCLUSION:
THE SUSTAINABILITY OF COPING STRATEGIES

To sustain coping strategies, these groups may benefit from some outside assistance in the form of material inputs and equipment for production, technical advice, credit or grants where necessary, and organizational and management training.

Although most of the activities that were identified above were initiated by the local people, there has been some support from both inside and outside the District from the government, churches, NGOs and foreign government agencies. Most of the soil and water conservation activities have received help from the MIDP. Other agencies which have assisted include: UNICEF, SIDA of Sweden, the Danish Volunteer Service and the Nordic Co-operative Project. In 1986–7, the MIDP assisted 25 SHGs in the District with materials worth Kshs 6,875 per group and twelve W/Gs with materials worth Kshs 4,000 per group.

During 1986–7, people's contributions and efforts attracted government and NGOs assistance amounting to Kshs 3.8 million. The NGOs contribution was Kshs 3.6 million, while the central and local governments respectively contributed Kshs 155,387 and Kshs 64,370. This shows great financial reliance on NGOs which are generally funded from overseas. This trend is likely to continue for some time, as many foreign-funded NGOs as well as donor agencies are increasingly interested in funding local groups in the country. This should be encouraged, provided that it does not destroy local initiatives. Such contributions and assistance add substantially to the sustainability of these micro-level activities.

Most of these activities are locally initiated and implemented although some external support has been received. The groups should be encouraged to be self-reliant by making financial and material contributions for themselves; only after this should they be considered for outside financial help. Some groups which seem to be successful in our sample, such as the Kamoli Ka Irongom W/G in Kibwezi and YSWGED in Yatta, are those that have made self-contributions and have then approached NGOs and governments for help. The groups that are mainly involved in socio-ecological activities should be encouraged to diversify and to include economic activities which will enhance the group's survival and self-reliance.

The groups that were studied did face varied obstacles and problems in attaining their objectives. Problems included: financial mismanagement by their leaders; inadequate financial support and insufficient inputs of new materials and equipment; lack of elementary skills by a majority of members, including high illiteracy rates; poor marketing channels; and inadequate infrastructure. These difficulties require active and flexible responses in a timely manner by the government and by NGOs and other agencies working in the area. This support should be given in a way that does not discourage these micro-level initiatives but instead enhances their development.

The NGOs and other donor agencies should continue to support these activities, especially through provision of technical and financial information and infrastructural support as well as management and organization training, since they are better placed to do so. The government should provide the overall stimulative policy framework. Regional or rural planners need to take into account the activities of local people and to integrate their initiatives and activities when planning for these areas. They must also allow local people to participate fully in the planning stages.

ACKNOWLEDGEMENT

I should like to acknowledge the support of the United Nations Centre for Regional Development under whose auspices some of the data used in this chapter were collected.

HOUSEHOLD BASED TREE PLANTING ACTIVITIES FOR FUELWOOD SUPPLY IN RURAL KENYA

The role of the Kenya Woodfuel Development Programme

Noel A. Chavangi

BACKGROUND TO THE KENYA WOODFUEL DEVELOPMENT PROGRAMME

Both the Kenya National Energy Symposium, held in November 1978, and the International Workshop on future energy policy issues for Kenya, held in May 1979, highlighted the need for Kenya to address the issue of woodfuel scarcity in the country and the problems of ensuring its future sustainable supply. At present, fuelwood accounts for 53 per cent of the total energy consumption in Kenya and is dominant in the rural areas (which use up 72 per cent of the total fuelwood consumption in Kenya). Charcoal accounts for 6 per cent of the total energy consumption in Kenya and is dominant in the urban centres (which use up 50 per cent of total charcoal consumption in Kenya). Fuelwood is mainly produced and consumed in the agricultural areas of high and medium potential, while charcoal is produced in the semi-arid areas and consumed in the urban centres (O'Keefe *et al.* 1984).

In view of the crisis in the woodfuel energy sector, the Beijer Institute of the Royal Swedish Academy of Sciences was invited in 1980 to assist in carrying out a system-wide analysis of the woodfuel supply situation in Kenya, which would form the basis for future programmes of energy provision. As a result of the analysis, a woodfuel supply strategy was outlined in the 1984–9 development plan of Kenya (Kenya 1986). It stresses the need for agroforestry programmes, peri-urban plantations, industrial plantations, and active participation of the Ministry of Energy (MOE) in continuing rural afforestation and soil conservation programmes.

The Kenya Woodfuel Development Programme (KWDP) is one of the projects operating under the auspices of the Ministry of Energy. The overall objective of the KWDP arises from one of the main findings of the study, to the effect that woodfuel (fuelwood and charcoal), as both a non-commercial and commercial source of energy, accounts for about three-quarters of the total energy base of Kenya and that the scope for employing other locally based energy sources is limited. In addition, the economic trend is unlikely to encourage a move from woodfuel to other fuels (oil or coal). As a consequence, woodfuel will continue to play a dominant role within the energy economy of Kenya (O'Keefe *et al*. 1984).

The greatest proportion (80 per cent) of Kenyans live in the agricultural lands of high and medium potential, which have a high degree of land demarcation and privatization and therefore offer almost no possibilities for the promotion of communal fuelwood plantations (Bradley 1984). Efforts towards the supply of energy for the millions of Kenyans living in these areas thus have to concentrate on the promotion of increased tree planting activities on individual household land (Van Gelder and Kerkhof 1984). The KWDP has worked with farmers in Kakamega and Kisii Districts of Kenya and in the process has gained useful experience in efforts towards developing viable agroforestry practices which are aimed at increasing the fuelwood supply to rural households. The objective in this chapter is to indicate the role that the KWDP has played in establishing, with farmers, self-sustaining innovations that would result in increased tree planting activities for fuelwood supply in rural areas. A particular concern is to focus on the way in which the KWDP has confronted male control of tree resources, in a context where women are responsible for gathering fuelwood for their energy needs. In the first section of the chapter, the philosophy and objectives of the KWDP are outlined. The case study of KWDP activities in Kakamega District which follows indicates the response of communities to the Programme.

The KWDP

The KWDP is basically a research and development programme whose objectives centre on the concept of sustainability. To achieve

149

this end, its concerns are: to develop replicable and participatory methodologies through which locally feasible agroforestry recommendation and locally tailored extension approaches and methods can be designed to ensure the effectiveness of woodfuel development at the district level; to verify both the feasibility of agroforestry recommendations and the effectiveness of extension approaches/methods, through which rural households can be encouraged to increase their woodfuel production for self-consumption and for the market and to economize on fuel consumption and end-uses; to assist in the development of Kenyan capacity to prepare, execute and monitor district fuelwood projects; and to develop a monitoring methodology at both district and national levels.

To achieve these objectives, a sequential implementation strategy covering several phases is adopted. First, a district resource analysis is carried out. In the initial phase the natural and human resources are thoroughly studied and analysed. District-level information which is derived from aerial photographs is combined with documentary information on population, soils, rainfall and land tenure. Analysis of these data indicates the extent of ecological zones, community differences and, hence, the selection of representative communities. From this base, particular households are selected for the field-based investigations in an effort to define more specifically the woodfuel problem and to lead to the design of interventions in subsequent phases (Bradley 1984). The strength of this approach emanates from the development research concept of involving the local communities in project identification and design.

In the second phase, the agroforestry and cultural backgrounds of the people are determined and defined. The information obtained contributes towards the design of the interventions that accommodate indigenous expertise and practices. The KWDP strives from the outset to determine who the target group is, what it needs and how the people themselves perceive the potentials for and constraints to achieving a desirable solution that is acceptable and appropriate to each cultural setting (Chavangi et al. 1986). After field trials of technical options, the KWDP launches a series of training programmes for extension staff of agencies which are already operating in the districts. This ensures that the tree planting messages are fully integrated within already existing extension services and that information is relayed to a larger community after the KWDP activities have been phased out of a district.

KAKAMEGA: A CASE STUDY IN KWDP ACTIVITIES

Kakamega District

Kakamega District, which has been the focus of KWDP activities since 1984, is representative of the densely populated, high-potential agricultural areas of Kenya. It has population densities, which rise to 1,000 inhabitants per km^2, and small farms averaging under one hectare in size, resulting in very heavy pressure on the land for food and fuelwood. Most of the land has been consolidated and registered in individual title. Little communal land remains. The degree of involvement of farmers in tree-raising activities is high, resulting in significant woody biomass coverage on the farms, with a large pro-portion of the trees having been planted by the farmers themselves. The extent of the farmers' existing knowledge on tree-raising techniques is great, giving a valuable reserve of indigenous experience which had hitherto been much underestimated and thus remained unexplored.

The KWDP survey, the main source of data for the following dis-cussion, showed that, in Kakamega District as a whole, the area that is covered by trees and shrubs on farms in rural areas increases as the population density increases and as average farm size decreases. The proportion of on-farm woody biomass which is deliberately planted and managed increases similarly. Approximately 20–25 per cent of the area of farms is covered by trees and shrubs of which about 75 per cent have been deliberately planted. As many as 79 per cent of farmers plant seedlings or sow seed directly; about 38 per cent of farmers raise seedlings in small (on-farm) nurseries, with the number of seedlings raised varying from a few hundred to over a thousand.

Owing to the high degree of land demarcation and privatization tree-raising activities are very much confined within each household. Opportunities for establishing communal fuelwood tree plantations are almost non-existent, so there is no doubt that any efforts to introduce a programme with the ultimate aim of providing a self-sustaining supply of fuelwood has to concentrate on the promotion of tree planting for fuelwood on individual farms. The survey results gave firm grounds for believing that the observed fuelwood shortage is the result not of a shortage of woody biomass on individual farms but of social and cultural forces within the households which deter-mine control over and access to the wood produced.

It was found that trees are planted and managed by farmers in

151

Kakamega District for many purposes: poles or timber for construction: split wood for sale; wood for making charcoal; and trees for use in rituals and religious ceremonies. There is a District preference for exotic tree species, such as eucalyptus, cypress and pine, which are regarded as cash crops or as a form of investment to pay for school fees. Indigenous tree species are rarely planted because they are thought to be able to grow on their own. In only a few areas are indigenous tree species, such as *Sesbania sesban*, deliberately grown on agricultural land. Sesbania is generally regarded not as a tree but as a plant which enhances soil fertility (by nitrogen fixation) and so is tolerated by farmers. This species has been effectively adopted as a source of fuelwood by the women in these areas. The KWDP is adopting this line of approach by introducing similar tree species (to Sesbania) which can serve the same purpose but which bypass the cultural blockage.

Considerable variations over small areas were found to exist both in the tree configurations (woodlots, hedges, trees within agricultural land) that are found on the farms and in the tree species (mainly exotic) that are grown within them. Tree production activities and preferences are to a very large extent locally determined. Farmers exchange seeds and/or seedlings with relatives and neighbours within their immediate surroundings in preference to relying on centralized nurseries which may be too far away to have any direct influence. The diffusion of innovative practices or new species throughout a community of farmers (as occurred with the spread of *Eucalyptus saligna* and other exotic species in Kenya) takes place gradually and cautiously at first, based on personal interactions.

A cultural survey revealed that only men have permanent rights to ownership of land. Trees are often planted to demarcate farm boundaries and thus serve to demonstrate ownership. Since trees are regarded as permanent features of the landscape, only men are able to plant them and to have sole rights to the use of them. A woman does have certain rights of access to her husband's land but not to trees deliberately planted on it. As a consequence, women do not enjoy the freedom to participate effectively in tree-raising activities. However, it is the woman's responsibility to provide for the household fuelwood needs. Fuelwood has until recently been regarded as a common and free good that may be collected from areas of 'natural vegetation' near or on the farms in the District. However, these sources have dwindled, primarily as a result of increasing population pressure, and women now have to walk longer distances in search of fuelwood or have to raise funds to purchase fuelwood.

Thus, fuelwood procurement was identified specifically as a woman's task. The KWDP considered that it would therefore seem logical to work through women in efforts to increase the supply of fuelwood to the rural households. However, the existence of myths (taboos) that influence the behaviour of women became a major obstacle to introducing a project aimed at increasing the participation of women in tree planting. To overcome this, an awareness programme has been developed which exposes the issue within the community so that it can be discussed and pragmatic solutions sought. The mass awareness activities, which are based on popular theatre (drama), are aimed at the community and the focus is on providing not only a mirror (to stimulate reflection) but also desirable images to catch the people's imagination and thus stimulate action-oriented discussions. The drama has now been developed into films for ease of handling. The awareness programme focuses on encouraging tree planting for fuelwood as a new concept or innovation since, hitherto, trees were rarely planted for fuelwood although fuelwood occasionally has been obtained as a by-product.

Farmers' tree planting activities

In each subregion or zone, there are a locally preferred mix of elements (woodlots, hedges, wind-rows, trees on agricultural land) and tree-raising activities, depending on local perception and practices regarding the agricultural production system. Some types of tree planting configurations are found on virtually all farms, i.e. trees in cropland, and trees in home compounds and hedges.

On-farm agroforestry practices

Investigations which were carried out to assess the actual practices by which the on-farm woody biomass is established revealed the importance of planting new species in the District. It was estimated that in the District over 42 per cent of the farmers had planted trees on their farms during the previous year. It is also recognized that the bulk of these newly planted trees come not from official nurseries, such as those of the government, forestry department or chiefs, but from the farmers' own nurseries or neighbours' on-farm nurseries. There exists a free sharing of seedlings between neighbours. Direct sowing

153

of tree seed and the collection of wildlings are also widespread practices in tree regeneration efforts on the farms.

Species preferences

The seedlings that are most commonly planted are exotic. In Kakamega, 85 per cent of the seedlings raised, bought or obtained for free were exotic tree species (mainly *Eucalyptus saligna* and to a lesser extent *Cupressus lusitanica*), whereas, of the wildings collected, 52 per cent were indigenous species and 39 per cent were fruit trees. About half of the trees sown directly were exotic species, mainly *Acacia mearnsii* (Black Wattle), *Eucalyptus saligna*, *Cassia siamea* and some indigenous species, mainly *Sesbania sesban* and *Croton microstachys*, in some parts of the District (Van Gelder and Kerkhof 1984).

Tree ownership uses

Trees are planted for various reasons across the District but mainly for construction poles and timber and as a form of investment, to be sold for cash. Such activities are under the control of men. Trees are also planted as a legal requirement for land demarcation purposes, hence the presence of hedges (as boundary features) on virtually all farms in the District. In Kakamega District, the planting of trees is mostly done by the men. As trees are generally viewed as men's property, men decide what to plant, where to plant and when to harvest. This is seen to be in line with the fact that, in most societies, men are the legal owners of the land and therefore have control over most of the farm resources.

Access to and control of household resources

In general, men and women have differential access to resources at the household level. Men, in their capacity as traditional heads of households, are usually in control of major household resources of economic importance. This control, as has been shown above, encompasses trees and hence men are found, in general, to control the planting and managing of trees on the farm, especially in the communities where traditions are observed to be relatively strong. The current trend in farm forestry activities focuses mainly on trees as a form of family investment (cash crop concept) and tends to overlook the requirement for a rational approach to total household tree and tree product needs. Thus, it is found that the heavy involvement of men, to the exclusion of women, also means low priority for women's needs for tree-related resources (firewood, fodder, raw materials for crafts, medicinal herbs) in relation to seemingly more pressing needs of the household. The service role of trees in agroforestry systems, including enhancing the production of organic matter, maintaining soil fertility, improving soil structure and thus reducing soil erosion, are all factors which are of major importance in subsistence agriculture (mainly done by women) and which are usually overlooked. To achieve a balance in providing for the total household woody biomass requirements, most of the issues raised above have to be considered and tackled at the household level. The focal point is the need to foster increased involvement of women in tree planting activities at the household level, in efforts towards providing for women's specific needs, enhancing their status, and thus increasing their self-confidence and pride. There are certain tasks that have been associated with women since time immemorial. Women still feel a sense of fulfilment in being able to meet the demands of these tasks. For rural Kenya, the provision of fuelwood is one such task.

KWDP INTERVENTION STRATEGY

Equipped with the knowledge that farmers in Kakamega have built up a large body of knowledge of agroforestry practices that are applicable to their particular farms, the KWDP mapped out intervention strategies that involved: creating an awareness of the fuelwood issue as a family problem to be tackled by both men and women; increasing dialogue within the households on fuelwood-related issues and initiating discussions towards common solutions; increasing the scope of tree planting by women through the introduction of the concept of multipurpose tree species, which can be managed by women; promoting woody species which are relatively fast growing and which do not yet have a cash value attached to them. These species are planted specifically for fuelwood but they also improve soil structure and soil fertility because of their nitrogen-fixing ability. In view of small farm sizes, multipurpose species are essential. A series of interrelated agroforestry and extension activities for the implementation of the overall programme was developed. These activities included: on-farm trials with individual farmers; the group extension approach; and a mass media approach, involving use of drama, films, printed pamphlets and brochures and radio messages.

On-farm trials

Initially the programme carried out reconnaissance research covering just over 520 farmers in Kakamega. In Kakamega, 28 farms were selected for on-farm trials, on the basis of farm size and gender of the farm manager. Farm sizes varied from 0.5 ha to 7 ha. Of the 28 farms, three belong to widows, five belong to women whose husbands work away from the farm (female-headed households because of migration), and 20 belong to married couples, both of whom reside on the farm. Farmers were provided with a few seedlings of quick-growing fuelwood species and small amounts of seed. Each farmer received 15–20 seedlings of three species and 50 seeds of one of the species. Four species were used: *Calliandra calothysrus, Leucaena leucocephala, Mimosa scabrella* and *Gliricidia sepium.* The KWDP was particularly interested in what farmers would do with the new materials on the bases of their existing knowledge of tree growing.

The choice of these four species was greatly influenced by the observed practice of women planting Sesbania in Ebusikhale in Kakamega. Sesbania has been accepted as a fuelwood species that women are allowed to plant and harvest. This species is locally considered to be a shrub, rather than a proper tree. Consequently, planting of this species does not establish any right to land. Furthermore, the species is not considered to be commercially valuable.

The farmers were given no technical information but the KWDP made arrangements for close observation of what the farmers did and discussed the activities with them constantly. The on-farm trials aimed: to create an environment for interacting with several farmers and building a relationship of mutual trust; to learn as much as possible of their knowledge on tree-growing and management (this approach was intended to expose the KWDP extension workers to the concept of learning from and with the farmers); to provide for individual farm visits to elicit information from farmers to identify problems; and to facilitate, from the outset, the participation of the farmers in finding solutions to the identified problems.

The KWDP staff members, therefore, decided to experiment with other designs with the farmer in order to take into account the farmer's knowledge and experience. The approach relied on mutual agreement as to where to plant the trees and, later on, to establish appropriate harvesting regimes. Together, extensionists and farmers have formulated ideas as to the most appropriate intervention strategies.

Although it is acknowledged that the small size of the trials dictates that the information be treated cautiously, there are a few worthwhile conclusions that emerge from a critical assessment of the trial on 28 farms in Kakamega.

The on-farm trials were used to initiate dialogue with the farming community. All efforts were geared towards understanding the community, analysing the fuelwood situation as perceived by the community, and stimulating the people into identifying possible solutions to the problem. Many of the options and questions that were raised through these farm trials have continuously been used in developing technical and extension options for the KWDP and in setting up the trials. The specific issues that were identified for further experimentation on a wide basis, in efforts towards assessing possible viable options, included: planting sites; spacing within the different configurations; performance of the various species; tree management practices for various end-use purposes such as fuelwood, fodder, soil conservation; compatibility with crops; resistance to insect and disease attacks; and seed handling and storage. The farmers in turn have become aware that they can solve part, if not all, of their fuelwood problems by raising trees which are fast maturing and which provide an opening to the cultural barriers that inhibit women's participation in agroforestry activities (Chavangi and Ngugi 1987).

One of the KWDP's objectives was to discover how local knowledge systems about trees would influence farmers' use of seed and seedlings. Survival of the planted seedlings was very impressive. On the whole their performance was observed to be encouraging, although not all did well (namely *Gliricidia sepium*), as many different planting strategies were used. An interesting range of tree planting configurations and sites was observed: the trees were planted in hedges, on agricultural land, near homesteads, on terraces, in tiny woodlots and as scattered trees all over the farms. Sometimes the trees were planted in pastures, and this led to destruction by animals if the trees were not protected. This problem is an off-shoot of the fact that three of the species that were used in the trial are palatable to animals. From comments by farmers, it would appear that they have intentions of planting the fuelwood trees on favoured land, which is a promising possibility (Chavangi and Ngugi 1987).

Harvesting of the trees had been sporadic. In most cases, farmers appeared to be waiting to see how big the trees would grow. On six farms where trees had been harvested by the end of two years, they were mainly used for fuelwood; indeed, the need for fuel was often identified as the reason for harvesting. Other uses were as rails for the construction of granaries and as supports for banana trees. The use of the foliage from the trial species for fodder was also observed. Those who had used the species for fuel indicated their satisfaction with the fuel qualities of the species, to the extent that those who wanted to continue growing the fast maturing species would do so for fuel in most cases; fodder and building materials were also mentioned as end-uses. On most of the farms, seed had been harvested from the trial species by the end of two years of growth.

No clear harvest pattern emerged from the 28 trials but obviously the number of farms in the trials was too small for such a pattern to emerge. It does appear, however, that the KWDP needs to support on-station experimental data with a wider body of data from farmers, a process which is already being undertaken.

Although most of the farmers had harvested seed within the first two years, only two farmers had established a nursery after harvesting the seed. There was some concern about the reasons why the farmers did not establish nurseries, direct sow and coppice regularly at a higher level. Why do the farmers keep seed and where and how do they store it? Why do they prefer giving seed away? While it is possible that they do intend to establish nurseries in the future, only about three said that they would do so. Perhaps the KWDP has expected too much in too short a period of time. There is a need to communicate with the farmers regarding the coppicing qualities of Calliandra in particular, and direct sowing for other species. It would appear that a majority of farmers are not aware of the coppicing characteristics; they also have not tried planting cuttings.

Although direct sowing was tried, it has been too limited. Only two farmers directly sowed the seed. Direct sowing is a convenient and easy method of regenerating some of the species, and it is assumed that the farmers thought of trying this out because they had previously done so for species such as *Markhamia platycalyx*. Collection of wildlings should also be encouraged so that it becomes widespread and intensified. Collection of wildlings has also been undertaken by the farmers, and some farmers are experimenting with coppicing of the species at a regular interval.

There seems to be no pattern in how the trial farmers perceive the species. Some species have drawn considerable comment while others have been mentioned only once or twice in passing. Leucaena, for example, has been noticeable for its horizontal growth and slowness in increasing in diameter and for its prolific seeding. Calliandra, on the other hand, has often been identified for its ornamental characteristics, while Sesbania appears to have been easily acceptable but not particularly noticeable. Not many comments have been forwarded about Mimosa but, amongst all of the trials, there has been a general impression that the farmers, especially the men, have been watching Mimosa with keen interest. This is suspected to have something to do with *Mimosa scabrella* showing signs of growing into a big rounded tree, in comparison to the other species, which are shrubby.

From the discussions with the farmers, it appears that the species have raised great interest among the neighbouring farmers. In fact, in a majority of cases, the seed that was harvested by the farmers had been given away to neighbours, friends and relatives. Judging by the high incidence of this, it would appear that the species are diffusing rapidly amongst the farmers. Some farmers also indicated that their neighbours had been asking questions about the trees.

Measurement of attitudes or perceptions is normally difficult since these are hard to gauge from brief interviews with a respondent. Also, attitudes and perceptions may change from time to time, or even from place to place, depending on the surroundings and context of a discussion. Nevertheless, long term interaction with an individual or individuals can lead to an understanding of their attitudes, whether or not these alter. In the case of the KWDP's trials with 28 farmers, it has been possible to gain a general view of what the farmers really think of the trial species, and how they perceive them in terms of their possible contribution to the household energy budget and other farm needs.

The issue of how ownership affected management and harvesting was broached in the structured discussions which took place on a number of farms. Some women claimed ownership of the trees but on most of the farms the trees were reported to be owned by 'the family'; a few women considered the trees as belonging to their husbands, even though the husbands do not live on the farms and did not participate in raising the trees. There was no distinct difference among farms in terms of who cared for the trees: on most farms, the husband and the wife or son cared for the trees jointly. In other cases, it was reported that the whole family looked after the trees. In a few cases, a woman cared for the trees on her own.

In some cases, women indicated an intention to regenerate more trees in the future for fuel and other needs, such as the building of granaries. Where the men expressed an intention to regenerate more of the species, the intended end-uses were more ambivalent; a list of uses, such as construction poles, railing and fencing materials, was given, sometimes with reference to particular species such as Calliandra, Mimosa and Sesbania. Fuel was also mentioned by men but it was difficult to ascertain the priority that they assign to different uses. A few men indicated that they wanted to grow trees for fuel for their wives, which in itself is a promising development.

It was apparent that, barely two years after the trials began, some farmers had already given up on the species. In one particular case, a woman who was interviewed indicated her husband's dissatisfaction with the species and stated that they had no intention of regenerating any more trees. She was not willing to express her own opinion on the species but claimed that her husband, who had expected to be given Eucalyptus and Cypress, had received only fuelwood species which he did not believe were worth growing.

Male ownership of the trees appears to be a prominent feature. In those cases where the trees had not been harvested at all, the most common owner was a man. Nevertheless, there are indications that some men have already left the job of caring for and harvesting trees to their wives, while others do it for them. It would appear that some men are keen on leaving the species to grow for a longer period of time in order to observe them. So far, male ownership does not appear to have affected adversely the orginal intention of promoting cultivation of trees for fuelwood, since the majority of harvested trees have been utilized for this purpose.

Generally, interplanting of the trees with crops was observed to be an acceptable practice. Some women have been impressed by the fact that, having been unable to grow *Eucalyptus saligna* (a common species in Kakamega), they have now found some species which they can plant in cropland. Judging from the indications on planting site preferences, it is also possible that the farmers will in future experiment with the trees in a variety of planting sites. However, it is recognized that a farmer, in order to make meaningful decisions on the most productive planting site, needs to grow many more trees than the small number given to the 28 farmers.

The KWDP has realized the importance of explaining to farmers the growth characteristics of the trial species and their 'shrub-like' features, since unrealized expectations could cause adverse reaction to the fuelwood species. Such explanations help to avoid disappointment on the part of the farmers at the poor performance of the trees. Those who had been expecting the trial species to grow into big, thick trees were disappointed when they realized that this would not be the case for some of the species.

The group extension approach

During November 1984, the KWDP decided that the trials on the 28 farms were sufficiently promising that larger scale, on-farm trials of the same species could be implemented, based on the group extension approach. Interested neighbours of the initial 28 farmers were brought together in groups. Where it was not possible to form a group on this basis, existing organizations were used, including a women's group, a church group and an adult literacy class. Also, one youth and one men-only group were formed.

In total, 25 groups were formed in Kakamega with 520 members. The majority of group members (75 per cent) were women. The average size of each group was 25, with members basically from the immediate neighbourhood and drawn from the same social class. Most participants were from the middle economic level of the community, rather than the very poor or the well-off.

The extension staff held, on average, three separate discussions per group, generally at two-week intervals. Through problem-solving tactics, the group members were assisted in analysing the fuelwood situation, defining the problem and coming up with possible solutions to the identified fuelwood problem. The group members were then encouraged to work out a course of action which was aimed at meeting their needs. Any external assistance which was needed was also determined.

The overall strategy, apart from learning about the possible motivation components around which an effective awareness programme could be designed, was to learn about various technical practices of farmers and the adaptability of the selected fuelwood species. Areas of specific emphasis were: species performance; spacing trials; tree management practices for various end-use purposes; planting sites; regeneration methods; compatibility with crops; resistance to insect attacks; and alternative uses, such as fodder, soil conservation and soil fertility improvement. Seed handling and storage were also monitored.

The few seedlings trials that were conducted all over the District, using the 28 farmers, had raised a lot of interest. Believing that such interest could be indicative of a positive demand for seed, group members, assisted by project staff, decided to establish Seed Production Units (SPUs) which could also be used as demonstration plots. The functions were specified as follows: to monitor seed potentials of the new species; to function ultimately as seed orchards for the community; to demonstrate the growth performance and characteristics of the new species and to promote community interest and stimulate discussions; to facilitate choice of species once an individual had decided to try planting of introduced species; and to involve school children, both girls and boys, in raising trees for fuel-wood, as a possible means of overcoming cultural constraints. Each group aimed at establishing one SPU but only 16 groups succeeded. Of these, 10 SPUs are located on farmers' lands, 5 are in church compounds, and 1 is on public land. Additionally, 14 SPUs were created on school compounds and 4 on public land.

A group activity such as this was greatly encouraged so as to promote and maintain group identity. Thus, the project promoted the establishment of an SPU in which all group members could actively participate during group meetings. In addition to the group activity, each member decided to plant trees on his or her land. Married women had to negotiate individually with their husbands while youths had to negotiate with their parents for land. The KWDP supplied members with seedlings. Each member decided on the number of seedlings, subject to a maximum of 250 seedlings, and chose the type of seedling to be supplied from a selection of four species. Slightly over 100,000 seedlings were distributed, with an average of 150 seedlings per farmer.

Group extension activities were monitored in May 1987, two years after seedling distribution. Out of the 520 farmers who participated in Kakamega, 109 respondents were interviewed. These 109 were randomly selected and therefore did not take into account the propor-tion of female-headed households. Of the 109 respondents, 83 per cent (96 respondents) had trees still surviving; 52 per cent (57 respondents) had harvested the trees. Of these 57 respondents, 65 per cent had harvested for fuelwood, 14 per cent for fodder and 14 per cent for other reasons, such as for construction rails.

Only some farmers planted these agroforestry species in their agri-cultural fields. No negative effects of the trees on the maize and beans crops were reported, and trees were considered to be compatible with crops. A few farmers reported increased maize yields during the second year (Chavangi 1988).

As indicated in Table 7.1, generally 30 per cent of those involved in tree-raising activities are women. However, women were responsible for 50 per cent of wildling collection and 80 per cent of direct seeding (Chavangi and Ngugi 1987).

The impact of the group extension approach comes out quite

Table 7.1 Summary of monitoring results of the 25 groups
in Kakamega District

Activity (%)	Husband	Wife	Son	Others[1]	Total (N)[2]
Tree ownership	65	30	<5	0	109
Tree management	44	32	13	11	109
Tree harvesting	60	30	8	2	57
Seed harvesting	44	15	15	26	26
Nursery establishment	37	0	37	24	18
Wildling collection	25	50	25	0	12
Direct seeding	—	80	20	0	5

1 Others include: daughters, in-laws, labourers, neighbours, friends, and other
relatives.
2 N = 109.

clearly in discussions with participants, both women and men. The
group extension efforts have been effective in improving the fuelwood
situation for individuals, as illustrated by the experience of Rispa
Akhoya from Eshianda sublocation of Marama Location, Kakamega.
Her church group, the Mother's Union, has participated in the
KWDP group approach extension efforts since 1984. She received
250 seedlings in early 1985 (March–April) and first harvested in
December of the same year. As she relates, involvement in the
KWDP activities has been a positive experience for her:

> I am glad that I participated in the activities of the programme
> right from the start and that I did not ignore them as I had done
> with earlier programmes, such as the Sugarcane and Poultry
> ones. My main fear was in regard to the credit element, in case
> my land could be auctioned if not successful.
>
> I am a Christian and so the cultural beliefs did not bother me
> very much. I attended the initial meetings, requested 250 seed-
> lings, and planted them in a woodlot and a few in the home
> compound. I am grateful for the encouragement I received
> from KWDP staff. My trees did well; I only lost a few from
> insect attacks.
>
> Before participating in the fuelwood production activities, I
> had a major fuelwood problem. But since I started harvesting
> the trees, I have enough fuelwood and I have even sold some. In
> 1986, I increased my woodlot with another 400 trees. I raised the
> seedlings in a small on-farm nursery using seed from my older
> trees. In 1987, I planted another 200 trees along the terraces.

I have been very happy with the programme activities. I have very good firewood, I have shared the seeds and wildlings with my neighbours, and I am encouraging more women to take up the fuelwood raising activity.

(Chavangi 1988)

From the experience and information gained from both the on-farm trials and the group extension activities in Kakamega, a number of conclusions were reached. First, the idea of tree planting for the specific supply of fuelwood has to be introduced in the communities as an innovation. Second, men have to plant these trees to assist their wives in producing fuelwood or they have to allow the women to plant and use the trees for fuel. Third, tree-raising for fuelwood touches on the culturally determined relationship between husband and wife and upon traditional tree and land ownership issues. Finally, an 'awareness raising' programme is needed, and forms the basis of the mass media extension approach.

The mass media approach

Both intensive work with individual farmers and the group extension approach have been conducted in only 11 out of 202 sublocations in Kakamega. To disseminate information more broadly and to encourage participation in tree-raising activities, the KWDP has launched a mass media programme. This programme first sought to raise awareness about fuelwood issues and to provoke discussions of problems and possible solutions. Later, the mass media messages presented more technical information on tree-raising. The KWDP has employed drama, films, public rallies (community and school based) and school field days to sensitize communities to fuelwood issues. These public events have been supplemented by printed extension materials, such as a picture-story (comic book), whose written messages focus on tree-raising techniques. Radio messages have been utilized also. Messages have been transmitted in serial form (drama, film, radio or pamphlets) to provide 'boosters' to maintain farmers' interest. To ensure that issues which are raised by the mass media efforts take root and are sustained in the community, the KWDP relies on the farmer-to-farmer extension approach that is supported by follow-up activities with groups and individual farmers.

One drama production was pre-tested in November 1985 and shown in January/February 1986 at 14 different sites. A total of

16,500 people (7,700 adults and 8,800 children) attended the drama. One week later follow-up meetings were held with interested farmers to discuss the issues raised and to give demonstrations on nursery establishment, root pruning and other techniques. On these occasions, seed was distributed to 3,630 farmers. Three months later, in May 1986, 777 of the 3,630 farmers were interviewed to monitor the results. It was found that 73 per cent of those interviewed had established nurseries with the seed taken.

Although the play was well received, the KWDP decided to replace it with a series of films. Problems had arisen in performing the drama as a result of illness or lateness on the part of the actors. Also, the logistical problems that were involved in transporting a cast of over twenty individuals were considerable. Films were shown in 18 different sites during July and August 1987. Out of 24,747 attending the shows, 70 per cent were children, 15 per cent youths, 8 per cent adult females, and 6 per cent adult males. The opportunity was taken to distribute 10,966 seed packages.

The KWDP extension staff noted a generally favourable audience reaction to the films. As illustrated by comments expressed in Murhanda sublocation, the films generated much discussion:

'It's good to have brought the film during the day so that we can also see it.' (Female adult)
'My home is the same as that one, only I cannot break a chair or the house. In fact, I am afraid I don't know what to use when the rains come. Ask whether they have those trees to give us.' (Female adult)
'Can women really plant trees?' (Female adult)
'I saw a woman in the film buying firewood. How many things shall I be buying in my life? Food, clothes, fees, and also firewood? No! I will plant my own trees for firewood!' (Male adult)
'This depicts our daily lives. It comes from nowhere but our homes.' (Female adult)

(Chavangi 1988)

In November 1987 a monitoring exercise was undertaken to assess what farmers did with the seed that they obtained after the film shows. On average, 84 per cent of those who took seed had established on-farm nurseries.

The guiding principle for future work is based on the understanding that the Ministry of Energy has no intention of establishing a parallel extension system to those already existing within the collaborating

165

ministries. Therefore, the energy message to the farmer has to be integrated within already established extension agencies. In line with this, activities which are planned for the next three years focus on the integration of fuelwood energy plans with other rural development activities. For the planned activities to begin smoothly, a workshop is viewed as important for coming up with recommendations as to how to proceed in efforts towards establishing mutual linkages among activities.

OVERALL CONCLUSIONS

Work with farmers in the on-field trials, the group extension approach and some initial monitoring of results from the dramas and films have led the KWDP to a number of conclusions. The initial problem that was identified in Kakamega, namely that domestic woodfuel shortages are caused, not by technical problems, such as lack of knowledge of the benefits of planting trees, but by socio-cultural constraints, was correct. Individual farm visits have proved invaluable for learning from and with the farmer. These visits have been especially useful for eliciting information regarding the identification of problems and for stimulating experimentation.

The group extension approach has proven effective in verifying the KWDP's understanding of the community's perspective, in formulating a description of the system, in helping to establish priorities, in setting up workshops and seminars, and in evaluating and pre-screening ideas. The initial focus on introducing as an innovation the idea of tree planting for the specific supply of fuelwood has been effective in sensitizing farmers to the need for increased tree planting activities. The end-use, however, is determined by the most pressing need at the time of harvest. Thus, the multipurpose species that were introduced offer the greatest long term flexibility.

Since the programme has undertaken several activities to promote women's participation in tree-raising, it is difficult to assess the impact of each activity separately. Overall, it seems that efforts to promote community and household discussions of fuelwood issues have been most significant. All of these efforts – working with individual farmers, groups, and the larger community through mass media approaches – have encouraged the planting of trees for fuel-wood by both women and men. Where cultural beliefs concerning the division of labour by sex are still strong, extension approaches focusing on working with the household as a unit are proving effective.

The focus on planting quick-growing, multipurpose agroforestry tree species has encouraged the participation of women. These species are slowly being accepted as fuelwood species. By offering alternatives that do not clash with existing patterns of tree use, such as the use of timber as cash crops and as a means of establishing claims to land ownership, the eventual result should be a greater control of fuelwood trees by women.

Since the KWDP activities took place in only a small proportion of District sublocations, it is considered too early to assess whether or not the overall fuelwood situation has improved. Of course those farmers who have participated in growing fuelwood species for at least two years, a majority (20 out of the 28 in the on-farm trials and 65 per cent of group members interviewed) of those harvesting trees have used the wood for fuel. This information suggests that the fuelwood situation for participating households has begun to improve. It should be noted, however, that the number of seedlings involved was small (up to a maximum of 250 seedlings per household). Preliminary estimates indicate that an average household of six persons would require approximately 1,000 trees to attain self-sufficiency.

REMAINING QUESTIONS

First, for any innovation to take root, people must be able to observe benefits from their efforts. Further assessment is needed of self-sufficiency in fuelwood on the small farms using on-field trials. Further experimentation is needed on the technical performance of the species and on the fuelwood quality in order to formulate technical recommendations, including issues such as planting sites, spacing, regeneration methods and management practices for various end-use purposes.

Second, little research has been carried out on indigenous tree species that can contribute to the fuelwood supply. The effect of establishing an economic value for tree products, especially seed, must be examined in order to determine whether this can promote interest in fuelwood species. Third, promoting tree-raising which is less labour-demanding must be further assessed as to whether such practices can effectively encourage women's participation.

Finally, it has not been determined whether direct promotion of tree planting for fuelwood encourages increased planting of indigenous tree species for fuelwood supplies. If this should prove to be the case, then planting trees for fuelwood will have become established practice.

SUMMARY

For the KWDP activities carried out in Kakamega, the Initial District Resource Analysis has proved invaluable (Bradley 1984). This analysis helped in identifying the main fuelwood shortage indicators, in understanding the fuelwood problems as perceived by the communities, and in designing a realistic woodfuel implementation strategy.

The KWDP has utilized two broad approaches in order to improve the fuelwood situation in Kakamega. First, there has been a focus on initiating dialogue among women and men at the community level in the hope that such dialogue filters through to the household level. This approach was necessary in Kakamega because women, recognizing that fuelwood procurement was their task, believed themselves to have failed if they sought their husbands' assistance. On the other hand, men did not want to be associated with the fuelwood issue because it fell outside their mandate. Thus, the provision of fuelwood had to be recognized as a task which, owing to changing circumstances, needed the co-operation of husband and wife. It was necessary to bring the two parties together to discuss the issue, so that there would be a recognition that fuelwood could no longer be handled by women alone and that men's involvement was necessary.

Second, once solutions were agreed upon and action taken, the availability of seed of the quick-growing, multipurpose agroforestry species provided possibilities for the direct participation of women. The species were considered to be shrubs and not trees, and so the cultural taboos limiting women's participation in tree-raising activities could be overcome. The two approaches, therefore, were complementary. Given the situation in Kakamega, one could not have succeeded in the absence of the other.

However, planting of multipurpose tree species for increasing fuelwood supplies to rural households must be seen as an introductory effort. Communities are starting to think of fuelwood as a resource which can no longer be obtained for free but which has to be either produced or purchased. The fuelwood development efforts, therefore, have to be viewed as supplementary to other efforts which are designed to increase the availability of forest resources, such as the supply of fuelwood as a by-product on the farm and the collection of fuelwood by women.

To increase fuelwood supplies where strong cultural divisions of labour are found, the strategy of working with households and encouraging the participation of men, women and children seems to work better than focusing on the women only. The guiding principle here is for trees to be planted for fuelwood regardless of who actually does the planting. If this is achieved, then the provision of fuelwood is no longer an issue specific to women.

The mass media approach has proved effective for the purpose of sensitizing communities to the fuelwood problems, of stimulating discussions, of raising interest and of initiating action. To sustain interest and maintain action, support through the other extension methods, such as approaches involving both, is necessary. This ensures the achievement of visible results that can initiate the diffusion process.

8

LOCAL COPING STRATEGIES IN DODOMA DISTRICT, TANZANIA

Japheth M.M. Ndaro

INTRODUCTION

This is a case study of the local or community level strategies that people have devised and adopted in response to, and as a means of coping with, the recent economic crisis on the one hand and the hostile ecological conditions on the other. The economic crisis has arisen as a consequence of the interplay of two types of factors: internal and external factors. The internal factors that have consistently affected economic performance in Tanzania include:

1 Insufficient resources for agricultural development compared with the high priority given to industry. Between 1976 and 1986, the agricultural sector received only 13.54 per cent of capital investment compared to 24.20 per cent allocated to industry (United Republic of Tanzania 1986).
2 Inadequate producer incentives and marketing/distribution systems. Since the mid-1970s, the real producer price index of all export crops has fallen consistently. Between 1982–3 and 1984–5, real producer prices were increased by 5 per cent per year but producer prices in real terms are still only half of what they were in the early 1970s (Odunga *et al.* 1988). On the distribution front, the main problem to be faced is that the country's physical infrastructure has been deteriorating for want of essential repairs and maintenance. Such deterioration has inhibited the timely movement of agricultural products and inputs to the rural areas.
3 Expansionary fiscal and monetary policies which added to inflationary pressures and thus to exchange rate over-valuation. Economic growth in Tanzania has been characterized by high inflation in recent years. The rate of inflation in 1984 accelerated to

over 36 per cent from an annual average of 11 per cent during most of the 1970s, as measured by the National Price Index.

4 Unfavourable weather conditions. Until the early 1970s, Tanzania was largely self-sufficient in the major food staples: maize, paddy, sorghum and cassava. This situation abruptly changed in 1973 when drought struck many parts of the country. Following this situation, over the last two decades, Tanzania's performance in agricultural production has varied considerably from severe food crises, such as in 1974–5 and again in 1982–3 to 1984–5, to comfortable surpluses, as in 1986–7 and 1987–8.

5 Domestic economic dislocation, in particular foreign exchange shortages affecting the supply of fertilizers, agro-chemicals, farm equipment, incentive goods, and constraining crop processing and transportation. The scarcity of foreign exchange considerably reduced the country's ability to import machinery and other inputs for most sectors in the economy, and caused a sharp decline in the quality of essential services, particularly health and education.

External factors have also contributed to the general economic decline. The budgetary situation deteriorated sharply after Tanzania's unaided effort in Amin's war of aggression in 1978–9 and the effects of the break-up of the East African Community. The 1979 increase in petroleum prices added US\$ 150 million to the annual oil import bill. This increased to US\$ 180 million in 1986 (United Republic of Tanzania 1986). Fuel shortages in recent years have had a crippling effect on economic activity and on the quality of life throughout the country. Meanwhile, a world recession and declining commodity export prices caused a major deterioration in the terms of trade. The trade deficit worsened as a result of the 1980–1 drought which necessitated large imports of food grains (Odunga *et al.* 1988).

The combined effect of these factors has, over time, greatly reduced the ability of the government to meet most local needs satisfactorily. The shortage of foreign exchange has led to underutilization of capacity in industrial establishments as a result of the inability of the government to import spare parts, raw materials and new machinery. As a consequence, a prolonged shortage of industrial incentive goods, such as clothes, sugar, cigarettes, soap, and torch batteries, has prevailed in the country. Community services such as health and education have also been handicapped. Import requirements of medicines, medical equipment, training materials, laboratory equipment and chemicals cannot be sustained. Joseph

Warioba, the Prime Minister and First Vice-President, has put the argument more succinctly:

> although the Economic Recovery Programme shows positive signs, the country's economy is still bad. The economy still suffers from constraints such as undercapacity utilisation, unfulfilled agro-needs, crop haulage and storage. People in the villages experienced problems of unsold crops or lack of transport and high prices for inputs. In towns, workers also were facing problems as their income failed to meet basic needs due to continuous decline of the shilling. People should therefore join hands with the Government in undertaking different projects as the only sure way to better their lives.
>
> (*Daily News*, 3 April 1989)

The study area is Dodoma District, located on the Central plateau of Tanzania, some 500 km west of Dar es Salaam. Dodoma District has a dry, savanna-type of climate which is characterized by a long, dry season lasting between late April and early December and a short, wet season occurring in the remaining months. The District is one of the least developed areas in the country and is the poorest in terms of income per capita. The District economy is almost entirely dependent on arable farming and animal husbandry. Agriculture is characterized by low productivity, resulting from low erratic rainfall, high evapo-transpiration and low moisture-holding capacity of the soils. These conditions, together with widespread overstocking and overgrazing, make the District susceptible to extensive soil erosion and creeping desertification.

The unpredictable nature of the prevailing rainfall pattern has made the District prone to food deficits. Although the population has responded to this challenge by planting drought-resistant crops such as sorghum and millet, this strategy has not turned Dodoma into a food surplus district. Moreover, the prevailing rainfall pattern contributes significantly to the acuteness of the water supply problem in the District both for household and livestock use.

On the socioeconomic front, the existing economic crisis nationwide, as argued above, coupled with unreliable weather conditions in Dodoma District, has created unbearable living conditions. The hardest hit by the crisis are the peasants, the youth, women and low-income earners. These groups have been forced to seek additional or alternative employment to augment their incomes. Moreover, community services, such as health and education, have also been

172

handicapped by shortage of foreign exchange and the inability of the government to allocate adequate resources to these sectors. The movement of goods and services in rural areas has also been constrained by transportation bottle-necks in the economy.

The purpose of this study is to show how local populations have responded to the various problems discussed above in the light of examples taken from Dodoma District. The coping strategies which are outlined in the study have the objectives of meeting subsistence needs and of preventing a drastic fall in living standards of the local population in the face of the prevailing economic crisis. The strategies are implemented locally by means of small-scale activities which are geared to the capabilities of the individuals, groups and villages concerned.

PHASES OF COPING STRATEGIES IN DODOMA DISTRICT

In Table 8.1 a cross-section of local coping strategies in Dodoma District is presented. As the coping strategies were started in different time periods, a distinction is made between new and revived activities.

In Dodoma District, three phases of locally initiated coping strategies are clearly distinguishable. The first phase was the ten-year period after independence, 1961–70. During this period, the inhabitants of Dodoma devised and adopted strategies that did not conform with the political slogan of nation building which was dominant in the early 1960s. The strategies were concerned with trade and exchange, local crafts and industries, the construction of small dams and the opening up of new lands to facilitate shifting cultivation. They did not, for example, consider building schools or dispensaries or constructing roads. To a large extent, coping strategies in phase one were essentially survival strategies.

The second phase covered the period from 1971 to 1978 during which villagization campaigns were taking place across the country. This phase marked the abolition of locally based institutions, such as local authorities and co-operative societies, coupled with promises by the central government to meet expenses for local basic needs, in particular health, education and water. In concrete terms, the phase involved moving people in rural areas to specific locations in order to establish nucleated or *Ujamaa* villages where the government was to provide schools, health and water facilities. This arrangement quickly gained considerable goodwill among the people. In reality, however, it stifled local self-initiative in some parts of the District.

Table 8.1 Cross-section of coping strategies in Dodoma District

Major strategy or response	Activity undertaken	Main focus and objective of activity	Initiators	Contribution of activity	Degree of spread	Activity: new or revived
Economic	1 Local crafts and industries	Income generation	Individual artists	Production of household items, farm implements	Widespread among poorer people. Blacksmithing and metal trades involve mainly unemployed poorer youth groups.	Revived
	2 Exchange and barter trade	Increase food stocks for households	Individual households	Aversion of hunger and possible famine	Widespread among poorer people	Revived
	3 Charcoal making	Income generation	Individual peasants	To increase household incomes and provide energy to urban dwellers	Widespread among poorer people	New
	4 Selling animals	Income generation to pastoralists	Individual pastoralists	To provide income to pastoralists and manure/fertilizer to farmers to increase production	Widespread sale of manure largely by wealthy livestock owners	New
	5 Vegetable gardening	New employment and income generation	Individual youth and women's groups	Human resources mobilization	Moderate activity largely of middle-level income peasants	New
	6 Local beer making	Employment and income generation	Individuals and women's groups	To augment household incomes and employment	Widespread among poorer women	Revived

	Activity	Objective	Undertaken by	Purpose	Extent	Status
Economic (*continued*)	7 Opening up of kiosks	Employment	Individuals	Employment of youths	Moderate mainly among middle-level income peasants	New
	8 Door-to-door salespeople	Employment and income generation	Individual young men and women	Employment	Moderate	New
Social	1 Building dispensaries and clinics	To provide health care	Local village leaders	Ensuring a healthy population	Moderate	New
	2 Building primary schools and vocational centres	To provide primary and post-primary education for their children	Local village leaders	Raising the literacy level of the population and improving skills in various trades	Widespread	New
	3 Building of small dams	To provide water for household and livestock use	Local village leaders	Improving water supply system in villages during the dry season	Moderate	Revived
	4 Transportation	To provide rural transportation of inputs and outputs, and to facilitate human movement	Local village leaders	Crop haulage, human transport and incomes to village governments	Moderate	New
	5 Savings and credit societies	To provide cheap loans to members	Groups of people	Financial resource mobilization	Moderate largely among middle-level households	New

Table 8.1 Cross-section of coping strategies in Dodoma District (*continued*)

Major strategy or response	Activity undertaken	Main focus and objective of activity	Initiators	Contribution of activity	Degree of spread	Activity: new or revived
Social (*continued*)	6 Burial societies	To provide a social service to bereaved families	Groups of people in Dodoma urban	Social cohesion	Moderate largely among middle-level income groups	New
	7 Milling machinery	To provide milling services to villagers	Women's groups in villages	Lessening of the burden on women	Moderate	New
Ecological	1 Afforestation	Control of soil erosion and desertification	Local leaders	To prevent soil erosion and desertification and to provide soil cover	Widespread	New
	2 Destocking	Control of soil erosion	Local leaders	To prevent soil erosion, to improve the livestock herd	Low	New
	3 Opening up new farmlands	To increase agricultural output	Individual peasants	As a cushion against low productivity	Moderate	Revived

Source: From surveys conducted by the author between July and December 1988

In the third phase, after 1979, local self-initiative in development activities had virtually disappeared as people depended more and more on government handouts. At the same time, this phase coincided with general economic decline in the District. The occurrence of drought conditions in 1981–2 and again in 1983–4, accompanied by general price increases, shortage of industrial incentive goods and a rise in the cost of living, pushed many people in the District almost to the point of destitution. This situation left the population with no alternative other than to revive and intensify previous local initiatives and to contrive new ones. The range of coping strategies in force after 1979 is indicative of a dramatic increase in the number and types of self-help initiatives that have emerged or re-emerged in the wake of a deepening crisis in the district.

Clearly, coping strategies in Dodoma District have emerged in response to various historical developments. It is therefore important for our analysis to distinguish the circumstances in which each strategy emerged or re-emerged.

Phase I: 1961–70

An important characteristic of Dodoma District before and after independence has been the recurrence of drought and famine (Brooke 1967; Ndorobo 1973). Brooke has documented several drought years, beginning as early as 1919–20 up to and including 1961–3. This documentation highlights the fact that drought gave birth to famine and that people devised and adopted several strategies to deal with such situations. Among the coping strategies that were adopted, barter and exchange, and migration to other parts of the country were the most prominent. In other situations, people learned to store food in a common pool (normally near the residences of their chiefs) during good years and would use the grain stock in years of shortage. In the case of barter and exchange, people traded livestock for grain or exchanged their labour for grain. Discussions with knowledgeable people in the District corroborate Brooke's documentation of this phenomenon.

With the increasingly unreliable rainfall conditions in the District, grain shortage has remained a major bottle-neck facing the people. Moreover, lower returns from agriculture resulting from low erratic rainfall, high evapo-transpiration and low moisture-holding capacity of the soils prevented households from satisfying their subsistence

177

needs even in years of adequate rainfall. The inhabitants of Dodoma responded by expanding the practice of shifting cultivation in their farming systems. Knowledgeable people in the village suggest that shifting cultivation became more widespread in the District in the 1950s and that, since then, peasants have opened up new farmland whenever the existing land became exhausted.

A related ecological problem that afflicted the rural population in Dodoma in this period was the lack of water for domestic and live-stock use during dry seasons. Although the post-independence government embarked on a rather ambitious programme to provide water to villagers by sinking boreholes and shallow wells and by con-structing dams at Hombolo, Ikowa, Buigiri and Matumbulu, these efforts benefited just a few villages. In the majority of the villages, water supply remained an acute problem. Consequently, as early as 1962, people in a number of villages initiated the construction of their own dams to alleviate the problem. Attempts by the villagers of Chololo and Nala are worth mentioning in this respect.

Local initiatives were also prominent in the crafts and industrial fields in the period under review. The remoteness of most villages in the District prevented peasants from acquiring modern industrial goods such as household utensils and farm implements. Local crafts and industries emerged in Dodoma not only as a reaction to the non-availability of industrial goods but also evolved as part of the *Wagogo* culture. Traditionally, crafts and industries were limited to specific clans and such specialization remained in force even in this period. Crafts industries were scattered throughout the District, but the scale of production of the implements was not large. Moreover, this activity facilitated barter trade among the people. It was only in the late 1960s that the produce from this activity found its way into modern exchange channels and this development motivated artisans to improve both the quality and quantity of their products. All in all, the Arusha Declaration (Nyerere 1967) and the *Ujamaa* policy (Nyerere 1968), which marked an important milestone in the development of the country as a whole, did not inspire the people of Dodoma to engage in development initiatives that were alien to their sociocultural environment.

Phase II: 1971–8

One of the outstanding developments in this period was the villagiza-tion of the District which involved moving people either to the location of existing social services, such as schools, dispensaries, water points, and trading settlements, or to new areas. The size and layout of the new villages were based on the economics of providing social services and not on production. Consequently, Dodoma villagers did not alter their pattern of cultivation, crops grown or life pattern. Thus, they remained vulnerable to the exigencies of weather, low productivity and possible stagnation. Instead of developing an attitude of self-reliance, villagization temporarily caused some villages to become more dependent on government than was formerly the case (Kauzeni 1988). Given this situation, the survival strategies that were started in the previous phase were continued, and sometimes intensified. Attempts to construct earthen dams were made in several villages: Gawaye (1975), Mbabala B (1978) and Ntyuka (1981). Villagers opened up new farmlands to augment food production and hence continued to practise shifting cultivation. Barter and exchange increased because of the social and economic dislocation arising from villagization.

A development that contributed to the emergence of new coping strategies in Dodoma District was the decision of the central government to transfer the capital from Dar es Salaam to Dodoma in 1972. This action brought about a very rapid increase in the population of Dodoma Town from an estimated 38,000 people in 1972 to an estimated 158,000 people in 1978 (United Republic of Tanzania 1982). This sudden increase in population created new and additional demands for various items that could be obtained locally. The most critical of these were charcoal, firewood, vegetables and animal manure. Most of the new migrants needed charcoal and firewood as energy while the few who opened up vineyards created a ready market for animal manure. This development motivated villagers living around Dodoma Town to engage in charcoal production and the sale of animal manure to town dwellers. The former activity was carried out mainly by poorer members of the community, the latter by wealthier livestock owners. In addition, the middle-level peasantry found a ready market for vegetables and responded by growing this crop more intensively, particularly during the dry season.

A further development during this period was the establishment of village co-operative shops. Village shops were established in Dodoma in 1976 following the Bihawana Declaration which made it compulsory for each village in the District to establish a shop. Villagers were motivated to found shops not only to obtain the needed shop services but also as an investment outlet for their savings. From this explanation, it is clear that the idea of starting village shops was initiated by government rather than by the people themselves. In fact, it was in the form of an 'operation'. People's initiatives were limited to the way in which such shops had to function: for example, the number and size of shares to be contributed, the type of commodities to be sold in the shop and the management of the shop.

A particularly vivid example of a locally motivated strategy in this period concerned the poor rural transportation system. Rural transportation in Dodoma District had been neglected by the central government for a long time. This neglect affected not only the movement of people but also the transportation of crops from production sites to marketing centres. The late delivery of agricultural inputs to rural areas also worried local leaders. As a consequence, between 1973 and 1978, villagers and their leaders took the initiative by purchasing vehicles on a self-help basis: some villages (Chamwino, Msanga, Kigwe, Mwitikira) bought lorries, other (Mlowa Bwawani, Mlowa Barabarani, Mpunguzi, Mvumi) bought tractors and a few (Mpunguzi, Hombolo Bwawani, Chamwino) purchased buses (Mulazi 1984).

Phase III: 1979

Underlying the social and economic performance of Dodoma District after 1978 was the nationwide economic decline arising from the interplay of factors, as outlined in the introductory section, and the peculiar ecological conditions of the District. The long drought that affected the District between 1980–1 and 1982–3 created unprecedented hardships for the people, particularly in rural areas. Thousands of tons of food were brought into the District by the government to avert possible famine. Because of the magnitude of the problem, the government directed wholesale institutions, such as the Regional Trading Company and the National Milling Corporation, to sell food and other essential goods to co-operative shops only. This action gave rise to the mushrooming of such shops at workplaces in Dodoma Town and to the opening up of women's shops in rural areas.

The food crisis added another burden to an already endemic problem of scarcity of industrial goods such as soap, salt, sugar, matches, torch batteries, clothes and cooking. Long queues of people waiting to purchase food and other essential commodities at designated co-operative shops were common in most parts of the District. From villages surrounding Dodoma Town, between 1982 and 1985, people commuted on a daily basis to the town to buy food and other items. As a result of the severity of the crisis, well placed town dwellers even managed to trade maize flour to villagers for cattle or goats.

On the social front, community services such as health and education faced fundamental problems. The government had invested heavily in social services, especially health, education and water supply, yet at the end of the 1970s none of these services could obtain the necessary operational support (Coulson 1982). Hospitals and dispensaries in the District found it difficult to secure the required medicinal supplies. In fact, until recently, more and more people have been paying for medical services in private hospitals or dispensaries. The illusion of free medical care is disappearing fast in Dodoma District. In some of the villages, the people no longer depend on modern medical services, relying instead on the services of herbalists.

Education has also suffered severely. In the majority of primary schools within the District, pupils have to sit on the floor as there are no desks. Even with the re-introduction of local authorities, schools are starved of funds to purchase stationery for the children. The buildings of most schools are in appalling condition because of inadequate maintenance. Moreover, the District economy is unable to absorb primary school leavers, many of whom migrate to urban areas to look for paid jobs only to increase the number of unemployed youths in towns. With regard to water supply, the situation has been very similar to other community services. Many villages, such as Nala and Zepisa, which had water supply systems, no longer enjoy the service because of the lack of spare parts to rehabilitate them.

In general, the increase in the cost of living arising from high rates of inflation, the scarcity of foreign exchange, the shortage of essential industrial incentive goods and the deterioration of community services have had the combined effect of reducing the standard of living and quality of life of the District's population. The picture that emerges after 1979 is one of a deepening economic crisis that significantly altered the lifestyle of the population in both rural and urban Dodoma. Trade liberalization and good weather conditions have eased some of the socioeconomic difficulties which were experienced after 1979, particularly after 1986 (Odunga *et al.* 1988). But this encouraging performance has become meaningful in Dodoma only after the people themselves had devised alternative means of survival.

Several such strategies were continued from previous phases. Examples are barter and exchange, shifting cultivation, local crafts and industries, and income-earning strategies such as beer making. Others included charcoal making, vegetable gardening, selling animal manure, and transportation. New activities that emerged in the wake of the deepening economic crisis in the District included: the formation of savings and credit societies, the emergence of burial societies and building co-operatives, the initiation of income-earning activities such as the opening up of kiosks, the mushrooming of door-to-door salespeople among the middle-level peasantry, and the formation of youth and women's groups as a means of pooling labour and other resources among poorer members of the community in order to obtain economies of scale in production.

To summarize, the people of Dodoma District have engaged in self-help activities of different types for a very long time. The range and diversity of these activities has increased and intensified over the years because of the social and economic upheavals that have confronted the District. For example, in the past, crafts and industries were confined to certain clans only. However, over time, this trade has spread to other people, particularly youths who have formed small-scale industrial groups and have improved the quality and quantity of their products. Even the marketing of the product is much more commercialized than previously. In addition, women's groups are playing an increasingly visible role in crisis response, whether the crisis arises from natural disasters such as drought or from development policies. For example, women's groups in the District are now involved in income-generating activities, such as vegetable shops and maize milling, to supplement family incomes. Individually, women brew beer to generate income.

In Phase I, it was found that the strategies that emerged were mainly directed to the subsistence needs of the population. However, in the most recent past, this concern has broadened to include self-help activities which are service-oriented. Communities have found it necessary to supplement government efforts in fields which were previously provided by the government, such as health, education, transportation, and the control of soil erosion and desertification. The initiatives by local communities are indicative of development 'from within' which requires government recognition and support. The intervention by rural communities has come at a time when government ability to marshal resources for development has deteriorated drastically. In the next section, a detailed description is provided of the interventions that were made by the rural communities in Dodoma District in the wake of the economic crisis.

COPING STRATEGIES AND RESPONSES

Among the coping strategies studied here, a distinction can be drawn between those which are intended to deal with the economic crisis (local crafts and industries, metal trading, exchange and barter system, charcoal making, selling animal manure, vegetable gardening, local beer making); those which arise from the failure of the government to deliver essential services to the population (housing and credit co-operatives to provide social services such as health, education, water, housing, transport, credit, and burial services, formation of clubs and groups such as youth and women's groups); and those which are intended to deal with the hostile environment (drought, low productivity, desertification, soil erosion).

Economic strategies

The widespread development of economic coping strategies and responses in Dodoma District has come about in the wake of the economic crisis confronting the local population. This development is expected to produce multiple benefits for the local population:

(a) The generation of additional income to households;
(b) The creation of employment for the population throughout the year;
(c) Increasing food stocks in food-deficit households.

Local crafts

Handicrafts are a long-established local occupation in Dodoma District. The craft persons are private artists who are located all over the District and who are engaged in trades such as pottery, and in the manufacturing of wooden trays, carvings, mortars and stools. They have few tools and usually use a tree or a simple hut for a workshop.

Pottery is almost completely carried out by women. The product is a popular cooking utensil which is used in most rural households and in some urban households. With the existing high prices of industrially produced cooking utensils, locally produced pots are cheap substitutes for rural households and low-income earners in urban areas.

The manufacture of wooden trays, carvings, mortars and stools is undertaken exclusively by men. The items produced are household necessities in rural communities. There are no specialized groups of people involved in these crafts. The items are simple implements which do not require sophisticated technology in their production. Moreover, because of the numbers of artisans involved and the simplicity of the tools used, the scale of production for each artisan is small. Some of the items find their way into the Dodoma urban market and therefore earn income for the artists.

Blacksmiths and metal tradesmen

Blacksmiths provide an important service to farmers by manufacturing farming tools such as cutlasses, hoes and harvesting tools. This trade was neglected in the District development strategy for a long time. In the past, blacksmiths and metal tradesmen have operated in Dodoma on either an individual or family basis, their scale of production remaining small. Recently, youths have organized themselves into co-operative groups with the purpose of increasing the scale of operation by taking advantage of organized marketing.

Three co-operative groups were studied during the initial survey. The first is the Nzuguni Youth Group which has 20 members. This group manufactures traditional knives, spears and hand hoes. The group is based at Nzuguni Village, some 10 km east of Dodoma Town. The second group is the Muungano Talawanda Group which has 50 members and is based in Dodoma Town. The group makes traditional metal items such as cutlasses, knives, hoes, and charcoal stoves. A third group, which is also based in Dodoma Town, is the Awamu Vijana Group. This group has 20 members and makes buckets, charcoal stoves and cooking utensils.

Exchange and barter trade

Unreliable rainfall has made Dodoma District an area prone to food deficits. Although the population has responded to this challenge by growing drought-resistant food crops, particularly sorghum and millet, this step has not made the District a food surplus area. Exchange and barter trade are the main strategies that continue to provide relief for a large section of the rural population. Barter trade in Dodoma District is a common feature in the exchange relations between November and March each year. In the main, the exchange is between food grains and livestock. The survey revealed that, in the 1987/8 farming season, two tins of maize flour were being exchanged for one goat. The survey also revealed that in certain cases grain is exchanged for labour. This happens when a member of a household which is facing food shortage works on another family's farm for several days/weeks and is paid in kind. This practice is quite common among the *Wagogo* in Dodoma District and is known as *kuhemea*. Considering that Dodoma District is a very dry area and that for most years the majority of the rural population do not satisfy their sub-sistence needs for five months in any one year, barter trade is an important strategy for alleviating hunger and human suffering which is caused by intermittent unfavourable weather conditions.

Charcoal making

This activity is more widespread in villages surrounding Dodoma Town. A study by Graham Thiele has shown that charcoal making is most popular in those villages which are less than 24.3 km from the town centre (Thiele 1984). Participants, who are from poor house-holds, are mainly young men, although in certain instances middle-aged men are also involved. Most of the participating population undertake this activity on a part-time basis in the farming season but are fully engaged in the dry season, that is, from May to November of each year.

Charcoal is normally sold to Dodoma Town residents. The survey showed that the price of charcoal varies from season to season. In the farming season, prices are normally higher than average, indicating that fewer people are involved in charcoal making and that household food stocks in the villages are lower than the required level. It is common in Dodoma Town to see charcoal traders buying grain or maize flour after selling their charcoal. Moreover, since about 90 per cent of Dodoma Town residents use charcoal as fuel, charcoal making is a response to the energy needs of the urban population.

Selling animal manure

The application of animal manure in farming is not a new develop-ment in Dodoma District. What is new is the realization by pastoral-ists that farmyard manure is a resource that can be used to augment their incomes. Before 1982, animal manure was given free to grape farmers and orchardists. However, with the increasing number of grape farmers in the District, wealthier livestock owning families started on their own initiative to sell animal manure to vineyard owners in 1982. With the escalation of prices for industrial fertilizer, farmyard manure is becoming not only an important substitute but also a popular fertilizer among grape farmers, vegetable gardeners and orchardists. With a large livestock population, the potential for expanding the use of animal manure in Dodoma District is substantial.

Vegetable gardening

Vegetable gardening is an off-farm seasonal activity that has been initiated by enterprising young men largely from the middle peasantry. The activity has gained momentum in the last five years in both scale and extent. It is an important activity in the dry season where small-scale irrigation is used and is more common in those villages where small dams have been constructed. In the case of seasonal rivers which dry up in the dry season, young men dig holes in the sand from the riverbed using shovels to obtain water for their small plots. Men from more than thirty villages in Dodoma District are actively engaged in this activity. A large part of the crop, predominantly tomatoes and onions, is sold to Dodoma Town residents and a small portion is exported to Dar es Salaam. Vegetable gardening enables an important section of the rural community to be actively involved in productive work throughout the year.

Beer making

Local beer making is not a new activity in Dodoma District but has become more popular as an income-earning activity among women from low-income groups and petty traders. More than 98 per cent of those who were interviewed indicated that their involvement in beer making was in response to the current economic difficulties facing them. In rural areas, this activity is found in almost every village. It is an important income-earning strategy after the harvest season; in the farming season, grains are normally in short supply and beer production is reduced. Moreover, village governments and local authorities normally forbid people from engaging in the activity at this time to ensure that villagers do not exhaust available grain. In urban Dodoma, this activity is undertaken throughout the year. The very high price of industrially produced beer has contributed significantly to the high demand for locally made beer in Dodoma Town. For example, by the end of 1988, the former was selling at $2.00 per litre compared to $0.33 per litre for the latter.

Social and political strategies

Social coping strategies in Dodoma District emerged as responses to unfulfilled government and political promises to deliver essential services such as health, education or water after the villagization campaign in 1976. In another development, the emergence of such local initiatives was pre-empted by government policies which advocated the doctrine of self-reliance as a follow-up to the Arusha Declaration (Nyerere 1967) and the post-villagization policy. This development is aimed at:

(a) Complementing government efforts towards their own development;
(b) Providing important services to local communities such as health care, education and transportation;
(c) Widening opportunities for individuals or organized groups and mobilizing capabilities and resources for common beliefs;
(d) Preventing a drastic fall in living standards for local communities and improving their quality of life.

Provision of health care

Health care is an essential service in any community. With the nucleation of the population after villagization in 1976, demand has increased but, over the years, has not been met because of the economic difficulties confronting the country since 1977. The government no longer commits enough resources to this sector. The enthusiasm that accompanied the provision of this service motivated village communities in Dodoma to initiate the construction of health facilities in their areas on a self-help basis. Village leaders first discuss the matter among themselves and then table the issue before the village community. When the idea is accepted, village leaders hold discussions with authorities to establish whether or not it falls within the government policy of service provision.

After the District authorities have accepted the village request, village leaders mobilize the community to construct the facility. The villagers' contribution is normally in the form of labour and money. Technical assistance in the building of a dispensary is normally provided by the Department of Community Development through its building brigade. In certain cases, building materials are obtained through assistance given by non-government organizations such as the Community Development Trust Fund of Tanzania. For example,

187

the dispensary-cum-clinic at Mahoma Makulu Village was built by way of self-help and received material assistance from the Dodoma Municipal Council and the Community Development Trust Fund of Tanzania. Several other villages in Dodoma District have constructed health facilities in their areas in this way.

Education

Subsequent to villagization in 1976 and the adoption of the policy of universal primary education in November 1977, each village was motivated to build its own primary school where such service was non-existent. The initiative normally came from village leaders and government officials at the district level.

Rural communities, in their attempt to provide this service for their children, have built classrooms and teachers' houses on a self-help basis. In some cases, entire primary schools have been built in this way while, in others, building materials have been provided by government and by some religious institutions which are based in the District. About 60 per cent of the primary schools existing in Dodoma District have been built through self-help.

Universal primary education has also created a problem in the form of youths completing primary education in large numbers. In Dodoma District in the last five years, only about 2 per cent annually of primary school leavers receive post-primary education. This means that 98 per cent of youths completing primary education join the pool of unemployed youths. Rural communities, with the help of local institutions and non-government organizations, such as Oxfam, World Vision and the Community Development Trust Fund (CDTF), have responded to this problem by building multipurpose community centres. In such centres at Mbabala B, Mpunguzi and Kigwe villages, youths train in skills such as carpentry, tailoring, blacksmithing and pottery (Hiari 1982).

Water supply

The problem of water for various uses in Dodoma District is critical because of the unreliable nature of rain or surface water. In the dry season, water availability is problematic not only for household use but also for livestock use. During this season, large herds of cattle are frequently moved long distances in search of water. The survey showed that pastoralists move their cattle over distances averaging up to 25 km in search of water and pasture.

For household use, villagers who are far away from reliable water sources, such as shallow wells or boreholes, frequently obtain their water by digging small wells in the dried river beds. The difficulties which are associated with this popular method of water harvesting indicate that the magnitude of the problem is not being adequately tackled. Several villages in Dodoma District on their own initiative now construct small dams to deal with the problem. Examples of such villages include Chololo (1962), Nàlà (1962), Gawaye (1975), Mbabala B (1978), Ntyuka (1981), and Ihumwa (1981). The main handicaps to the provision of water in this way are the lack of expertise, the limited technological base, and the lack of equipment and implements on the part of the villagers (Ndaro 1987). With assistance from government or NGOs, this method of water conservation would provide a more certain means of water harvesting in water-starved villages in the District.

Transportation

Over the last two decades, various initiatives have been taken by villagers to provide transportation in their areas on a self-help basis. The main initiative has come from local village leaders who have been inspired by political and government officials within the region. Villagers make their financial or equity contribution through village governments. The main support has come from the National Bank of Commerce (NBC) through its subsidiary, the Karadha Company Ltd, and the Co-operative and Rural Development Bank. Advisory and technical support has come mainly from the Co-operative Division at the District Headquarters.

The initiative has resulted in the acquisition of vehicles by villages. Since 1973 the distribution of ownership has been of the following order; 24 villages have acquired lorries, 5 villages have purchased tractors, while 3 own buses. The objective of the villages in acquiring vehicles is threefold: to provide rural transportation for agriculture inputs and farm produce; to facilitate human movement within the district; and to increase/expand village government coffers. The collective initiative by village communities to provide transport services for themselves is directed to finding a solution to the problem of rural transportation (Mulazi 1984).

Co-operative savings and credit societies

These are registered societies which are started through the initiative of the members themselves. Members are mainly employees in

189

different institutions in urban and rural areas. The main purposes of such societies are twofold: one is to provide cheap loans to members in case of problems such as bereavement, sickness, or school fees for their children; the other is to provide cheap loans to members for personal development, such as farm expansion, or for the purchase of important but rather expensive household items such as bicycles, radios or radio cassettes.

There are a total of sixteen such societies in Dodoma District. Three-quarters of the societies are located in Dodoma Town, while the remainder are found in rural areas of Dodoma District. The collective effort which is shown by such groups of people is designed to mobilize what little financial resources are available for their own development. Existing co-operative savings and credit societies in Dodoma have so far made it possible for members to engage in farm expansion, poultry projects, purchase of building materials, and other self-development activities.

Burial societies

Burial societies have been initiated by individuals from middle income households in urban areas for the purpose of assisting one another in cases of bereavement. Ten such societies are known to exist in Dodoma Town. Each burial society has a fund which is established through members' own contributions. The specific objectives of burial societies are:

(a) To provide financial and other material support to bereaved family members of the society;
(b) To assist in the transportation of the body of the deceased to the original place of domicile for burial;
(c) To provide assistance to members or travellers belonging to their region who happen to be in financial difficulties or need of material support.

Youth groups

In the last decade, several youth groups have emerged and formed co-operative ventures. These groups are of a social and economic nature. Two types of youth groups have emerged in Dodoma District, those which are spearheaded by religious institutions and those which are organized by the Department of Youth and Culture.

Among the former, two were visited during the survey: the Evangelical Youth Group is involved in pig and goat rearing projects in Hombolo village, while the Anglican Cathedral Group is engaged in vegetable gardening and fruit farming projects. In addition, this group cultivates drought-resistant crops such as sorghum and cassava. A total of 80 youths aged between 12 and 35 years are involved in these projects.

In the case of the youth groups that are organized by the Department of Youth and Culture, only one such group was visited. The Mpunguzi Youth Group, which is based at Mpunguzi Village some 21 km south of Dodoma Town, is involved in grape farming, an important cash crop in the district. The group consists of youths who have organized themselves into a farming brigade, after realizing their own common purpose of raising their incomes by pooling whatever resources they had, especially their labour.

Women's groups

Several types of women's groups can be identified: tailoring groups, co-operative shops, milling machine projects, and informal women's groups which are spearheaded by religious institutions. Although these groups may be perceived as responding to the prevailing socio-economic crisis in Dodoma District, they also represent an opportunity for women to acquire access to and control over resources in a society where these are frequently denied to them. In Wagogo society, control over land and livestock is overwhelmingly in male hands. Women's groups therefore provide women with a means to improve their access to resources without interfering with such rigid relations of production.

Tailoring groups

In the survey, six women's co-operative groups were visited. The women form their associations voluntarily, members contributing to a common fund in the initial stages of the association. After registration, assistance in the form of material support is sought from the Department of Community Development or some other donor agency. The women's group at Bihawana Village was assisted by a religious organization called World Vision.

191

Co-operative shops

These shops are spearheaded by the local branch of Umoja wa Wanawake wa Tanzania. Thus, the membership is confined to UWT members. There were more than fifteen women's co-operative shops run by the UWT in Dodoma District in 1984. However, by 1988 this number had declined to six. Factors which contributed to this decline included: very low initial capital; management problems arising from members' inexperience; and high operational costs.

Milling machine projects

Over the last decade, many women's groups in the villages, under the auspices of local UWT branches, have endeavoured to acquire milling machines in their villages. In certain villages, women's groups started other economic ventures, such as restaurants or co-operative farms, in order to raise funds for the purpose of acquiring milling machines. However, available evidence suggests that this method of raising funds has not been successful for many groups. Consequently, more effective methods have been sought. The most popular method has been to approach local non-government organizations, particularly the CDTF. In such situations, milling machines are delivered only after the relevant group has constructed a permanent structure to house the machine. Examples of women's groups which have acquired milling machines in this way are those in the villages of Mzakwe, Mlowa Barabarani, Nzuguni and Zepisa.

Informal women's groups

These are essentially voluntary groups which are initiated by religious institutions. The most active religious women's groups, which are found in Dodoma, belong to the Lutheran Church, the Anglican Church, the Roman Catholic Church and the Seventh Day Adventist Church. The groups are involved in tailoring, weaving and gardening. Initially, members contribute money to a common fund which is used to purchase the needed inputs. Material support is obtained from the relevant religious institutions. Proceeds from these activities are used by members to supplement their household incomes.

Ecological strategies

The strategies which are investigated in this category are devised in response to, and as a means of coping with, the hostile natural environment prevailing in Dodoma District. The strategies are directed towards the prevention of further degradation of the environment which is caused by soil erosion, deforestation and creeping desertification. The principal objective of these strategies is to improve environmental protection and living conditions through measures such as soil erosion control, afforestation, and desertification control.

Soil erosion control

In Dodoma District, soil erosion is caused by two main factors: overstocking, and deforestation as a result of charcoal making, house building and shifting cultivation. It follows that both pastoralists and agriculturalists contribute in one way or the other to soil erosion.

Responses or local initiatives from pastoralists and agriculturalists to soil erosion control are still neglible in Dodoma District. One possible approach to soil erosion control is destocking. In Dodoma, however, it is estimated that, since the introduction of such efforts in the mid-1970s, destocking has been insignificant (under 5 per cent) largely because pastoralists consider their livestock as capital investment. Pastoralists are not prepared to reduce the size of their livestock herds by selling or slaughtering them (Mascarenhas 1977). Village leaders, with the assistance of local district authorities, are attempting to circumvent this problem by designating specific areas as grazing land. It is thought that the restriction imposed on pastoralists by village government leaders will motivate them to reduce the size of their herds.

Deforestation and desertification control

Another approach to soil erosion control is afforestation or the planting of trees and the exercising of strict and conscious control in tree felling. In Dodoma District, there is considerable evidence to suggest that tree felling is undertaken indiscriminately, i.e. without any planned tree harvesting. Tree-cutting is undertaken either to enable peasants to acquire new farmland or to facilitate the production of charcoal and firewood for sale in Dodoma Town. Both practices are discouraged by the government as they constitute the main cause for the creeping desertification in Dodoma region.

193

Afforestation campaigns have taken place in villages and primary schools for several years now. In the process, village governments have designated specific areas for tree planting. During the survey it became clear that each village which was visited had established an average of three hectares of trees, which had been planted following the intervention of the village government. Primary schools had also planted an average of half a hectare at the time of the visit. In addition to these efforts, individual villagers are encouraged to plant trees near their households. Seedlings are normally supplied by local government authorities based at the District Headquarters. The survey also indicated that, so far, four villages have established their own nurseries to meet the ever increasing demand for seedlings in their areas.

IMPLICATIONS FOR RURAL/REGIONAL DEVELOPMENT

Local coping strategies are attempts by communities to provide essential goods and services in their own areas with little or no assistance from the state. The nature and scope of the activities that have been described in this study are manifestations of an approach to socioeconomic development which is grass-roots based. Specific strategies vary according to both socioeconomic stratum and gender.

The local provision of goods and services on a self-help basis, rather than a reliance on government assistance, is an important feature of a locally based rural/regional development paradigm. However, in the context of rural/regional development, there is a problem in that many local initiatives tend to be ignored in the main-stream planning process. In practice, such initiatives tend to have certain characteristics in common: they are often badly designed; they expect a great deal to be achieved in a short time from the use of meagre resources; they contribute little to raising the consciousness of local communities; they include activities which are marginal (in terms of both sustainability and expansion); and they are part and parcel of rural/regional development plans which are formulated and implemented by government. One result is the perpetuation of marginal social and economic roles for local-level coping strategies, which acts as a constraint on the potential for local populations to promote their own development.

There is, of course, no blueprint for achieving rural/regional development but the integration and co-ordination of local coping strategies with government initiatives may help to start such a process.

This approach can be adopted by a committed government and by organizations with decision-making power and access to knowledge and resources that can be used by local populations in their struggle for improvement and self-sustenance.

CONCLUSION

Grass-roots involvement in self-help activities has a long history in Tanzania, which can be traced back to 1967 when the policy of socialism and self-reliance was elaborated in the Arusha Declaration (Nyerere 1967). However, this policy did not provide for sufficient prominence being given to local strategies in national, regional or even district development plans. Instead, government initiatives continued to be given more prominence in development plans. Even with decentralization (Nyerere 1972), it has not been possible for local initiatives to filter upwards. Self-help initiatives are widespread in Dodoma District and other parts of Tanzania, as indicated in the study. In spite of this, local populations have been handicapped by their restricted access to finance, technology, organizational and management skills, advisory and technical services *inter alia*. To produce sound results, local coping strategies must solve the problem of access to these essential inputs.

In some instances, outside promotion of local coping strategies has been effected in Dodoma District. For youth groups, this intervention has taken the form of the Youth Development Task Force (Akator 1988); in the case of women, Umoja wa Wanawake wa Tanzania has had some influence. While the degree of contact with locally based groups has varied, such bodies have exerted little effect on mainstream decision-making. Of course, they have served to channel funds for short term, small scale projects for youth and women's groups, but their impact on grass-roots development has tended to be slight. Although there appears to be a serious need to re-assess their role, this does not mean they should be dismantled. The challenge is to make them more effective. Members and staff need adequate training in organizational, leadership, planning and management skills. Such institutions can play a catalytic role, working with local government not only to review and improve people's programmes, in line with acceptable targets and commitment for grass-roots development, but also to participate in the monitoring and evaluation of overall programmes. Their influence on district or regional decisions is likely to depend on their links with the grass-roots, since

195

organizations with a substantial power base among the people are less easily ignored.

ACKNOWLEDGEMENT

I should like to acknowledge the support of the United Nations Centre for Regional Development under whose auspices some of the data used in this chapter were collected.

9

THE INFORMAL SECTOR
A strategy for survival in Tanzania
Alice Nkhoma-Wamunza

INTRODUCTION

In Tanzania, the participation of women in the informal sector must
be seen in relation to the deterioration of the subsistence economy. In
the rural areas, women constitute the majority of the population and
form 80 per cent of the agricultural producers but they are also the
most marginalized section of the population. Women's access to land,
credit, agricultural inputs and technical know-how is limited compared to that of men. Further, although women work so hard, they
are not the main beneficiaries of the products of their labour (Muro
1987). Women are rarely represented in public decision-making
bodies nor do they effectively participate in the decision-making
process as in other fora; in fact, they have limited control over factors
that are vital for their economic survival.

The continuing economic crisis in Tanzania has affected women
and children more than men since it has led to cuts in public spending
in areas which most concern women, such as social and health
services. For example, government expenditure on health declined
from 9 per cent of total expenditure in 1973–4 to 4.9 per cent in
1982–3 to 1985–6. Similarly, government expenditure on education
declined from 14 per cent during 1975–6 and 1977–8 to 9 per cent for
the 1982–3 and 1985–6 period. Primary education has been the most
seriously affected. Here government expenditure declined from a
high of 19.9 per cent of total capital expenditure in 1982 to 11.9 per
cent by 1986 (Havnevik *et al.* 1988). For a country that is trying to
eradicate illiteracy and to strengthen its human capacity in the
management of various sectors, these cuts have had a negative impact
on the effort to improve the capacity of national human resources.

Such cutbacks reflect the post-independence economic performance of Tanzania. Tanzania's economic growth rate in the 1960s and early 1970s ranged between 5 and 7 per cent annually. But from 1973 to 1978 the economy began to decline rapidly; GDP averaged only 1.5 per cent per annum. Between 1978 and 1985, real income stood at 2 per cent while GDP stood at 10 per cent below the 1978 levels. There was, however, a slight increase in the growth rate of 0.3 per cent between 1985 and 1986 to 0.6 per cent from 1986 to 1987 (Havnevik *et al.* 1988). The deteriorating economy has meant a dramatic reduction in real earnings. The value of the T Shs to US$ 1 has continued to decline from T Shs 17 in March 1985 to T Shs 145 to US$ 1 in November 1989. One result has been an increase in the number of women and children living below the poverty line. Between 1972 and 1984, the average real wage in industry declined by 70 per cent. The high rate of inflation and devaluation has meant an increase in the cost of living while low wages and poor employment opportunities face the majority of women who are unskilled. By 1980 women constituted only 15 per cent of all wage and salaried employment. As a result, more women have been forced to engage in income-generating activities in the informal sector in order to sustain their families and to increase their purchasing power.

As elsewhere in sub-Saharan Africa, the current economic crisis in Tanzania is characterized by low per capita incomes, low production, slow export growth, weak balance of payments, shortages of foreign exchange and essential goods as well as inadequate recurrent funds for the operation and maintenance of services. African countries have responded to the crisis by introducing various adjustment programmes in order to contain the situation. In the case of Tanzania, various structural adjustment programmes have been introduced, none of which has seriously taken into consideration the specific needs of women. Instead, the programmes have worsened the condition of women. The National Economic Survival Programme of 1981–2 focused mainly on the management of the external sector imbalance, increasing exports, marketing of food surplus, controlling government expenditure and producing essential goods for industry. Despite the good intentions of the programme, it did not address the specific needs of women. There was no mechanism in the programme through which women could have had access to the resources to improve production.

As the crisis deepened in the 1980s, it was felt that the National Economic Survival Programme was not adequate. Thus, the 1982–5 Structural Adjustment Programme was introduced. Like its predecessor, it did not attempt to address the specific needs and problems of women agricultural producers, such as issues of access to land or farm inputs, credit or division of labour. Instead, the removal of food subsidies and reduced support to social services negatively affected women and children. Trade liberalization made matters worse because few could afford to buy the goods that were imported.

Since the first two structural adjustment programmes failed to meet their objectives, the government introduced the 1986–9 Economic Recovery Programme (United Republic of Tanzania 1986) which aimed to increase both food and cash crops, to improve marketing structures and resources in the agricultural sector and to rehabilitate supportive production infrastructure. None of the three programmes gave due consideration to the vital role played by women in production, marketing or distribution.

The participation of women in the informal sector, in order to generate additional income, is not a new phenomenon for women in the urban and rural areas of Tanzania. However, the current economic crisis and the monetization of subsistence has created a high demand for cash to meet the rising costs of food items, clothing and other basic needs. This situation has negatively affected the nutritional welfare of children who are left in the care of either other children or ageing grandmothers. In many cases, children are forced to work late hours to help their mothers to generate income through petty trading; sometimes this results in their being absent from school.

Another factor which has led to women's increased involvement in the informal sector is male migration, especially from rural to urban areas. Migration has contributed significantly to the increase in female-headed households, which in turn has meant women taking initiatives to generate income through the informal sector to enable them to provide and manage family needs in the absence of their husbands. The informal sector, therefore, has provided an opportunity for women to earn independent incomes but it also has meant a shifting of family responsibilities on to women and additional workloads for the already over-burdened women. Since women's involvement in the informal sector is undertaken concurrently with other household chores and agricultural cash food crop production, this situation in turn has serious implications for the health and nutritional status of both the women and their children.

Cultural factors have also contributed to more women engaging in cash-earning activities, since women in many African societies are brought up to take responsibility for providing food for their families.

199

From early childhood a girl is taught the virtues of hard work. Parents will influence their sons in the choice of a spouse by emphasizing the qualities of hard work in a young woman. Girls grow up believing that food provision and household care are their sole responsibilities, thus furthering the unequal sexual division of labour.

It is not surprising that so many women are turning to income-generating activities in the informal sector in order to earn extra income and to supplement other sources of income for the continued sustainment of their families. It is no longer possible for low income and peasant households to subsist on the low wages and income from agricultural produce, the prices of which are determined in international markets. By taking on income-generating activities, women are increasing their workload and stress but they are also becoming innovative and are developing informal support systems as they struggle for the betterment of their families' livelihoods. For example, they draw on *Update*, a rotating savings and loans system, to which each makes a monthly contribution. In turn, a lump sum of money is allocated to each member for investment in a business or to purchase household items which would otherwise be beyond reach.

Women are also creating employment opportunities for other women. The major areas of concentration for women's activities are in food processing, bun and bread making, petty trade, poultry, pigs, pottery, dairy cattle, transport, handicrafts, tailoring and embroidery, food vending and brewing. According to women, the monthly earnings from these activities are much higher than the official minimum wage of T Shs 2,500 per month. The high incomes in this sector have attracted many women in formal wage employment, thus tripling the workload of most women. For example, a woman in formal wage employment may keep up to 400 poultry while women keeping poultry on a full time basis may keep between 500 and 5,000 poultry and realize more than T Shs 300,000 per year.

This chapter presents a case study of women brewers in Utengule Usangu, a *Ujamaa* village in the Mbeya Region. It is a study and history of women's struggles. It illustrates their strengths, weaknesses, aspirations and determination for survival despite the many obstacles and constraints. The struggles also show us lessons which can be learned and which perhaps can contribute towards developing alternative mechanisms to deal with similar problems. The material in this chapter is based on earlier fieldwork and research, sponsored by ILO Geneva, on employment opportunities for rural women in

Tanzania. The methodology that was used involved case history studies, participatory observation, meetings, discussions and interviews.

BREWING

For the majority of women in rural areas and for rural–urban women, brewing is not only a major contribution and source of income but also a lucrative business. Indulgence in brewing is an activity which women take on in addition to their other daily household and agricultural chores.

In Tanzania, women from different parts of the country brew different types of alcohol; for example, *komoni* and *kangara* which are made from grains such as maize, millet or sorghum, *mbege* which is made from unripe bananas, *bonasi* from ripe pineapples, *ulaka* from the cashewnut, *dengerua* from sugar cane and *wanzuki* from honey. While the above types of alcohol are controlled and prepared mainly by women, there are other types of brews of which men are the main producers. For example, men prepare *ulanzi* (bamboo) or *mnazi* (palm) wine. The actual preparation of these is not a demanding task; it does not involve cooking or the fetching of fuelwood or water as is the case in the brews that are prepared by women. The wines that men produce are tapped straight from palm trees or bamboo shoots. They are not labour-demanding, although they do require skills such as climbing the tall palm trees and tapping. In addition, whereas ordinary brewing is the domain of women, distilling spirits such as *gongo* is dominated by men. Distillation involves technological innovation with which men tend to be more conversant. However, more women are now learning the skills of brewing *gongo*.

Brewing is a demanding, labour-intensive activity requiring large quantities of water and fuel. For many women, the search for water means walking long distances to rivers, streams or wells. Even for those women with access to piped water, the presence of a water supply system is no guarantee that a regular supply of water will be maintained. Often many of these supply systems cease to function owing to a lack of spare parts, poor operation and maintenance and a lack of trained personnel to operate and maintain them. Where training has been given in operation or maintenance, those trained have always been men despite the fact that women are the drawers of water. Until recently, development aid programmes have not adequately addressed this issue when sponsoring village water projects.

For many households in Tanzania, fuelwood is the main source of energy for cooking and especially for the brewing process. The demand for fuelwood is so great that women now have to walk longer distances in search of fuelwood for both household consumption and petty trade activities, such as baking bread, making bricks and drying tobacco, since it is no longer possible to find fuelwood nearby. The government and various non-governmental and international organizations in Tanzania are just beginning to emphasize the importance of increasing local awareness of the hazards of environmental degradation and of the need to preserve and plant trees for the future supply and source of fuelwood.

The Utengule Usangu Brewers

Utengule Usangu is a small *Ujamaa* village in the Usangu plains in Mbeya Region. The village is about 60 km from Mbeya Town. By 1982 the village had a population of about 1,860 with a total of 340 households. The major economic activities are farming and animal husbandry. Rice, maize and sorghum are the major food and cash crops. Some cotton is also grown as a cash crop.

Beer brewing among the Wasangu (the people of Usangu) is as old as their history. It plays an important social, religious and economic role. In pre-colonial days, the presence of beer played a significant role in the organization and mobilization of labour for agricultural production or for any task that required significant labour power. However, during the colonial period, beer brewing became a new source of income to pay poll tax.

Today, income from brewing continues to be the major reliable source of income for many rural women and has made it possible for women to pay their share of the development levy that was introduced by the government in 1982. Every able-bodied individual over 18 years of age, employed or unemployed, man or woman, is expected to pay this levy to finance the local government authorities. For many rural women this has meant an additional financial burden. Often, women with a reliable source of income from brewing and petty trade have no choice but to pay the levy for their husbands who may not have the money at the necessary time. Men frequently rely on income from seasonal crop sales, payment for which is often deferred by crop buying authorities. Whether women in the rural areas should pay the development levy has been the subject of much debate in the country. In 1982–3, the village government collected T Shs 26,589 from villagers compared to T Shs 12,413 from market traders. This may be an indicator of the role of beer brewing in the economy of the village.

Brewing is predominantly a female occupation because the tasks involved fall within the gender division of labour for women, e.g. drawing water, collecting fuelwood, grinding or pounding, and cooking. Some of the drudgery in brewing has eased through access to technological innovations such as grinding machines. Today, women rarely use the mortar and pestle except when small amounts of grain need to be ground. At the time that this research was undertaken, the village had three grinding mills which liberated women from the tedious process of pounding the grain with a mortar and pestle. However, when there is no diesel or when the mills break down, especially when no spare parts are available, women must either resort to the traditional way of grinding or travel to another village with a milling machine. Most women prefer to walk the long distances to the next milling machine rather than use up their energy through tedious pounding.

The process of brewing is very demanding. Women wake up very early to light the fire, draw water and mix the ingredients. They must collect sufficient fuelwood. In some cases, women with means or younger women prefer to buy fuelwood from fellow women or men rather than fetch it themselves because of the long distances involved and to avoid having to carry a heavy load on their heads. In all of these activities women make use of their support systems to ease the drudgery and work burden, so that the drawing of water, the collection of fuelwood and the preparation process of brewing is often done with the help of friends. In some cases, child labour is utilized. This is noticed especially in the rural–urban periphery of Dar es Salaam where young girls help their mothers to sell drinks to customers.

These support systems/networks and the co-operation are necessary to ease the work burden since, apart from brewing, women must also perform all other daily chores. The support networks are not confined to the preparation of beer but extend to co-operative child care. Often a friend or relative will offer to look after another woman's children while the latter is busy brewing or selling.

Brewing is a technique and those without the specific knowledge or skill or those who would like to earn money but cannot brew for religious or other reasons will often ask a friend to brew for them as long as they provide the ingredients, water and fuelwood. Here support systems are used by women to enhance the economic capacity of fellow women.

The whole brewing process, including the boiling, mixing, cooling and serving stages, may take between eight and ten hours, spread over a period of one week. Meanwhile, preparations for the next brew are made so that immediately after one brewing is sold out, production of the next brew begins. How much and how often a woman brews depends on her need for cash and the time available. Some women may brew between 90 and 180 litres of beer per week. The price of a litre of beer varies between T Shs 20 and T Shs 30. In urban areas, the price ranges between T Shs 30 and T Shs 40 per litre. Earnings from brewing vary from individual to individual depending on the amount brewed. By 1982, T Shs 600 per person per brew was considered to be the highest, whereas recently women in villages in Mbamba Bay in Ruvuma Region indicated that they earned between T Shs 2,000 and T Shs 3,000 per person per brew. In Dar es Salaam, a woman may earn up to T Shs 6,000 per brew in a day, assuming that she does not sell on credit to some customers and that the quality is good. This difference in earnings can be explained by the rise in prices of food items and the devaluation of the Tanzanian shilling against the US dollar.

On face value, the earnings appear high. However, the money is spent quickly because of the prohibitively high costs of food and other essentials. In 1989 in Dar es Salaam, a kilo of meat cost up to T Shs 300 whereas in the rural areas a kilo of meat sold for about T Shs 150. A loaf of bread sold at T Shs 70. The official price of a kilo of sugar was T Shs 100 but often the price was inflated to T Shs 200 on the black market because unscrupulous entrepreneurs created deliberate shortages so that they could earn more on high demand goods. Sugar is currently one of those essential foods that is subject to government control. Only those shops designated by the government, such as the National Milling Mobile Shops, sell at the official price. Despite control, the *walanguzi* (unscrupulous entrepreneurs) sell it at an inflated price. Given the fluctuating prices, most women are unable to save; earnings go to purchase the highly priced raw materials. Even if they could save, the absence of accessible banking facilities is a problem. Those women who are not engaged in brewing earn a living by selling cooked food at public places or practise in other types of petty trade.

The social and economic backgrounds of the brewers differ, including teachers and nurses in addition to peasant women. Those in wage employment who participated in the study confirmed that they earned more from brewing than from their salaries. Brewers in

Utengule Usangu are of all ages; women from the age of 16 onwards engage in this activity. Young girls learn the art of brewing from their mothers at an early age. When older, they brew independently to earn cash. For the younger women, cash earned is spent mainly on material goods such as fashionable clothes. With older women, money goes into sustaining their families and extended families. Cash is needed for clothes, school uniforms, fees, basic essentials such as salt, sugar, soap, kerosene and medicine, in addition to amounts for hiring farm labour, tractors or ox-ploughs and carts for the transportation of agricultural products. It is also needed to pay their poll tax or development levy. In most urban areas, only older women who are married, divorced or widows brew beer.

Women have worked out their own mechanism of production control by brewing in turns so as not to saturate the market. Controlling the supply on the market ensures equitable distribution of income from brewing among themselves. Brewing is a risky business; if the brew turns out to be bad, it is thrown away and the woman loses the costs and the potential income of the inputs. Quality of brew is important if it is to sell.

The amount of brewing activity varies from season to season. Soon after harvest there is more brewing activity than during cultivation and harvesting time. Grain, such as maize or millet, is in abundance during harvest time and there is also plenty of money in circulation after the sale of the crops. By November, brewing activity slows down as women are busy tilling the land in preparation for planting and weeding. For some women the brewing business continues but on a smaller scale, while others brew in order to harness labour for agricultural production.

Both men and women indulge in drinking beer, either in small groups or as individuals. Opening hours for *vilabu* (s. *kilabu*) (beer halls) are 3.00 p.m. to 10.00 p.m., Monday–Saturday, and from 10.00 a.m. to 10.00 p.m. on Sunday. The hours of *vilabu* are deliberately controlled by the village government so that the more productive morning hours are spent in agricultural production. Women must brew sufficient quantities of beer to keep circulation of money within the village; if they do not, then their husbands will spend money on beer in the next village.

Since brewing is a lucrative business, more men are engaging in this trade than ever before. Because men have the time and the resources, they buy beer wholesale from women and resell it at slightly higher prices. Normally two or three men team up and buy a large quantity of beer from women at lower prices before it is taken to the beer clubs (*vikao*) (s. *kikao*). Then the men take the beer to small private rooms alongside the women's main beer club and sell it from there. Their pattern of selling is different, as all the men who wish to drink make a down payment as calculated by the businessmen. For example, a group of ten men may contribute T Shs 30 each for a 20 litre bucket of beer. The businessmen make a profit of T Shs 200, having bought the beer wholesale from the women at T Shs 100. In the end, the businessmen earn twice or three times more than the women. On the other hand, the idea of selling wholesale appeals to women, who are happy to be released from having to stay at the *vilabu* until the beer is sold out and so can spend the time on other chores. It is possible that women may be pushed out of actually selling beer and instead will remain at the periphery, enduring the drudgery of fetching wood and water and of coping with the practical aspects of brewing, while men assume the role of middlemen and control the beer trade.

Some women, however, have caught on to the new tactics that men are employing. They do not brew, but buy wholesale from fellow women and resell it to patrons in large quantities. In this way, they earn twice as much as they paid for the brew and have more time to carry out other household chores.

THE UTENGULE USANGU BEER PROJECT

This section will describe a case study of a women's co-operative venture in beer brewing, how they organized, what they achieved and the problems that they encountered in the process.

In 1969 a group of ten women came together and discussed the setting up of a beer club (Kilabu cha Akina Mama). The idea originated from some of the women who had accompanied their migrant husbands who worked on sisal estates in Tanga and Morogoro Regions. By setting up this co-operative venture the women were responding jointly to common problems which faced them within the community and the household. Their immediate goal was to generate enough money to buy a milling machine to ease the burden on themselves and other women of pounding grain. They also wanted to start a co-operative shop. The group thus began to organize and they elected a chairperson, secretary and treasurer. Any woman could join as long as she was willing to contribute labour for setting up a building, e.g. by collecting thatching grass and plastering walls. Each member was also expected to contribute T Shs 2. Initially the women set up business under a tree. Business was good, especially between June and October when there was plenty of grain available and more money in circulation. There were many potential customers as well, such as the cattle owners and other traders who came monthly to sell their products. The money that was collected by the women from sales went into improving their premises. By 1970 the women had constructed a complete new building and business was thriving. They were even able to hire the services of a watchman and a sweeper whom they paid from the proceeds of beer sales. They built two latrines on the premises. The women held regular meetings to review progress on their activities and to receive and discuss the treasurer's financial report. At these meetings they also formulated guiding rules on how to run the club. The women were happy with the progress made. They had organizing capabilities and were a highly motivated group. Business began to pay off.

However, this state of affairs did not last for long. With success came problems. The women began experiencing interference from male parties who had vested interests in the activities of the group. One businessman in particular tried to bring about confusion in order to break up the group. For example, he spread false allegations against the treasurer by stating that she was misusing the group's funds for her own gain. Members began to distrust each other. The group was also subjected to a lot of pressure from the village group leaders. The women were asked from time to time to make financial contributions for village activities which needed financing. The group was politically vulnerable and any attempts to protest were quashed with threats to close down the club. On other occasions, a health officer would be called in to condemn the premises as unhygienic, with threats that the club would be closed down unless the women paid a fine; the women always paid the fine. Such tactics employed by men were eroding the group's enthusiasm. Their confidence was also being shattered. The women were made to feel as if they had no right to organize and that they ought to be grateful as the village government was doing them a favour. Ultimately the group began to disintegrate.

Finally, in 1972, the women were dispossessed of their club on the pretext of mismanagement. The village leaders used their power to appropriate property which legally belonged to the women. But the motivating force came from the businessman who colluded with the village leaders. The club was transferred to the businessman. He took over from the women, without compensation, what they had collectively built up.

The women did not accept this takeover silently. They protested to the village government but received neither sympathy nor support, not surprising since the village government at that time comprised men only. Towards 1974, the businessman who had taken over the *kilabu* failed, in part because the women brewers organized collectively and resisted the takeover by deliberately refusing to supply him with beer. They still controlled the brewing because they made the beer in their homes. Instead of supplying beer to the man's business, they 'illegally' sold it in their homes or in other secluded places. The women thus reunited and gave each other moral support. The male drinkers too were supportive, as their allegiance lay with those who could supply them with the beer that they wanted. This silent and effective protest put the women back into business. In 1975 the *kilabu* was returned to them. Business began to flourish and by the end of 1976 they had managed to save enough money to buy a house from a farmer for T Shs 5,000. This was converted into a second *kilabu*. The women also paid for the renovations, building materials and two lamps.

The group was expanding; by 1976, there were about 40 members. Some women, especially those who brewed beer as a full time activity, became more successful than others. There were, however, organizational problems within the group. The treasurer and chairperson were frequently under attack for taking undemocratic decisions. They were accused of collaborating with the village leaders, since they made fiscal contributions to village activities without first consulting the group. Another of the organizational problems was the low level of training in book-keeping. Solidarity within the group began to deteriorate. In 1977, the village government called a meeting in the spirit of trying to assist the women to resolve their differences. The leaders advised the women to split into two groups and to share their assets. The group split into two and each group began to run its own *kilabu*. Later, each group elected its own office bearers.

The year 1978 was one of uncertainty for the women. The village leaders continued to harass them and to interfere with the groups' activities. One day, the two groups, who were summoned on short notice to a meeting organized by the village government, found themselves the centre of discussion. They were informed that the village government had decided to take over and run the two clubs as a village project. The women were also informed that both the watchman and the sweeper would now be paid by the village government. Although the women protested, they were told that the decision was final and that there would be no compensation for the appropriated property. Not all of the men supported the takeover but those who sided with the women could not effect a change. After the meeting, the two treasurers were summoned individually to the village office and each was asked to surrender whatever cash was in her possession. One handed over T Shs 600 while the other surrendered T Shs 1,400 in cash and each relinquished the inventory of her group's assets. The assets were also appropriated. Although the main reason that was advanced for the takeover of the clubs was the need for the village to have an income-generating project, other reasons were cited by both men and women in the village. One such was that the economic independence that the women had achieved threatened the men. This did appear to be the major reason for the takeover, since, during informal discussions with the village leaders, this author was told: 'walikuwa wanaringa sana hao wanawake na hela zao (the women had become big-headed)!' The village government could have initiated a project of its own but it was more interested in this project because it was lucrative and well established.

Since 1978 the project has become the major source of income for the village. With this income the government is able to employ a full time clerk of accounts, watchmen, herdsmen and a shop assistant. The control of beer brewing was such a decisive issue because it was not only lucrative but also a source of regular income. There is very little chance that the women will ever regain their clubs. The women did attempt to raise the issue of the hijacking by the village government with an official of the Tanzanian Women's Organization (Umoja wa Wanawake wa Tanzania), who paid a one day official visit to different villages in the area. At a public meeting she confronted the village government on the issue, charging them with taking advantage of their privileged positions. She appealed to the village government to return the clubs to the women. However, there was no follow-up to her visit and the issue has remained unresolved.

Organized as a co-operative, the women achieved unity, acknowledged their ability to run an independent economic venture and were able to make collective democratic decisions. They also benefited by owning property. Most importantly, they achieved economic independence and stability and the ability to decide how to spend their money, which in many cases went to sustaining their families.

Economic independence also gave them confidence, although this confidence was ultimately regarded as a threat by the men, who thought that they were losing control over their wives. This was a problem particularly for those men who had no regular source of income of their own. Economic independence also meant that the women had taken up an extra burden to support their families. The consequence of takeover of the project by the village government was that the women were not able to realize their original goals of buying a milling machine and of opening a co-operative shop with a view to expanding their business.

Women continue to brew individually but now more and more men are entering the trade as middlemen. They buy the beer wholesale from the women at the low price of T Shs 300 for 90 litres of beer and set up their own businesses of *vikao*. Here they reap high profits, since whoever joins a *kikao* pays a fixed rate in order to be a member. These middlemen earn up to four times as much from one *debe* (20 litres) of beer. Some women are now going into wholesale buying and setting up of their own *vikao*.

CONCLUSIONS

Economic ventures which are undertaken by women have many things in common. The activities are normally marginalized and they lack recognition from policy makers and sometimes from community leaders in the allocation of resources, such as access to credit, loans and land, even though the success of these ventures contributes to the sustainment of the family. Lack of training in project management and basic book-keeping skills affects the continuity of a project. Similarly, the absence of banking facilities within walking distance affects women's ability to reinvest and so strengthen their economic capacity. At times, when women have entrusted money for safe-keeping among themselves, to husbands or to the village government, they have found that the money has disappeared when they need it.

Poor leadership and lack of training in leadership skills and group management sometimes lead to mismanagement of resources, destruction of group dynamics, and vulnerability to manipulation by unscrupulous village leaders or other interested parties who may end up controlling the project or taking over from the women. Often, women's projects lack markets because women have no access to vital marketing information and marketing strategies or even to information on viable and profitable project ventures. This can also be linked with the high rate of illiteracy and the low levels of education and technical know-how. It is also true that, in rural areas, projects are often imposed upon women's groups by donors without the donors first consulting the women to find out what the women's priorities are.

Inadequate legislation has also had adverse effects on women, in that no protection for women's economic groups is ensured. An example is the Tanzanian Co-operative Societies Act of 1982 which was enacted to provide for the formation, constitution, legislation and functioning of co-operative societies as instruments for implementing the policy of socialism and self-reliance. Sections 22 and 23 of the Act provide for the formation and recognition of only one co-operative society in a village. Women's economic groups, which are often not part of the village co-operative society, are therefore not covered under this Act, leading them to be placed at a disadvantage. Even when women's economic groups can form a branch of the main co-operative society in a village, they cannot function entirely under the control of women since the society falls within the jurisdiction of the village government which is dominated by men. The Co-operative Societies Act does not affect women's groups in urban areas where the independent formation of women's groups is not controlled by legislation. A review of this Act would be in order so as to allow for the formation of independent co-operative societies in the rural areas.

In order to understand more fully women's economic and political situation, it is necessary to analyse customary law and property rights and the implications that these have for the rights and full participation of women. It is common knowledge in Tanzania that the majority of women in both urban and rural communities are governed by customary law which is overwhelmingly used to male advantage. However, with increasing conscientization some women are becoming aware of their legal rights. For example, there have been cases where women have sought court decisions on child custody and maintenance, on marriage and divorce and the division of matrimonial property. One of the landmark cases in matrimonial property in the Tanzania Court of Appeal concerns Civil Appeal Case No. 9 of 1983, involving Hawa Mohamed and Ally Sefu (Dar es Salaam Registry, unreported). The High Court held that child care and housework which is done by a wife should be regarded as contribution and joint effort towards the acquisition of property.

In this ruling the High Court set a precedent by recognizing the contributions that a wife makes in the home for the well-being of the family. This decision could not have happened if Hawa had not contested the division of property in court when the marriage broke down. What is also significant about this case is the recognition by some of the legal authorities that customary law has to change with time but that this can come about only if people, and especially women, initiate the changes that are required in collaboration with women's organizations, parliamentarians and the legal profession. Rutashobya (1988) has argued that:

> The customs and laws of our society can be changed to accord women the rights, respect and dignity which they rightly deserve if each one of us realizes that he or she has a duty to take part in building new social attitudes in which all people are regarded as of equal social worth so that the criteria for the allocation of private and public responsibilities is not sex but the individual's potential contribution.
>
> (Rutashobya 1988)

Various developments are taking place in Tanzania through the Law Review Commission. Among the laws being reviewed is the Law of Marriage Act No. 5 of 1971. Women have also expressed concern about the need to review the Customary Law Declaration Order No. 4 of 1963, the Employment (Amendment) Act No. 20 of 1975, the Income Tax Act No. 3 of 1973 and the Affiliation Ordinance Cap. 278 because 'provisions in these legislations deny women certain rights or adversely affect their interests' (Sinare 1988).

In addition to legal barriers, women's economic activities are often hampered by lack of adequate financial resources. There being no alternative, women tend to be satisfied with the little cash that they earn from an economic activity because of their pressing needs for cash to purchase the basic necessities. There is little scope for investment. Nevertheless, independent incomes give women a more self-sufficient stance in the family and it makes them more secure and less vulnerable to mistreatment by husbands.

The independence, security, confidence and self-respect that women achieve through economic activities sometimes have negative implications, since most men feel threatened because they do not like to lose their position of power in the family. Some men go to the extent of forbidding their wives from participating in economic ventures. Further, when women do achieve economic independence, men often retire from providing for the family and so the women are left with yet another burden of making ends meet to sustain the family while the men spend their income unilaterally on durable goods.

Within the context of the structural adjustment and economic recovery policies which have been introduced in Tanzania over the past decade, women's abilities to meet these responsibilities have been further strained. In these programmes, low priority has been given towards the allocation of resources which would otherwise raise the productive capacity of women. In addition, the low levels of education and of technical and managerial expertise of women and their lack of access to credit have meant that most women cannot own property or take up top management positions or similar jobs to enable them to exercise their capabilities in influencing decisions that affect their lives.

In the informal sector, women have found a venue through which they can interact and be exposed to political issues and economic practices. Participation in the informal sector has in turn contributed towards raising women's consciousness, as the case study on beer brewing in Utengule Usangu illustrates.

10

DEVELOPMENT FROM WITHIN AND SURVIVAL IN RURAL AFRICA

A synthesis of theory and practice

D.R.F. Taylor

INTRODUCTION

The purpose of this chapter is to draw on the empirical evidence that is presented in the case studies, together with other sources, and to attempt to point to new directions for both development theory and development practice in Africa. This linkage between theory and practice is important, although it was pointed out in the Preface that

> The validity of development approaches will not be determined as a result of theoretical and ideological debate, but in the realm of practice. The peasant families of Africa . . . are more likely to judge the validity of a strategy from its results rather than its ideological or methodological soundness.
>
> (Stöhr and Taylor 1981: 458)

Good practice must be based on an understanding of the theoretical underpinnings of the approach being used. Although difficult to define, 'development' is generally agreed to include a series of components such as increased economic growth, equity and distribution of the fruits of that growth, control of the population of its own destiny and the achievement of qualitative transcendental values. Development cannot be defined in purely quantitative terms and differs over both time and space. It is best defined in terms of the aspirations and values of people in their own social context and, in this sense, is probably only really meaningful at the sub-national scale.

THE CRISIS OF DEVELOPMENT IN AFRICA

Mackenzie analysed aspects of the major elements of crisis in Chapter 1. Although quantification at the macro level is fraught with difficulties, there is no doubt that at this level the quality of life for the majority of Africa's inhabitants has been declining in both absolute and relative terms. For many parts of Africa there is no compelling evidence that 'development', however defined, is taking place. Increasing degradation would be a better description than 'development' for the current trends. Adebayo Adedeji (1989: 33), the Secretary-General of the Economic Commission for Africa, has outlined the major features of this debilitating crisis which he sees as having three major manifestations:

(a) A deterioration in the main macroeconomic indicators;
(b) A disintegration of productive mechanisms and infrastructural facilities;
(c) An accelerating decline in social welfare.

To this list could be added an increasing deterioration of the physical environment. The figures that he quotes are disturbing, to say the least. Between 1980 and 1988, per capita income for Africa as a whole fell steadily by 2.6 per cent per annum and wage employment fell by 16 per cent. GDP per capita in 1978 was US$ 854; by 1988 it was US$ 565. In 1978 the per capita growth rate was 3.03 per cent; in 1988, it was − 0.88 per cent. Growth rates in all sectors are dropping and Africa's inflation rates are rising. The debt burden has risen from US$ 48.3 billion in 1978 to over US$ 230 billion in 1988 and debt servicing obligations now exceed 100 per cent of export earnings in several African countries. The scale of the debt servicing problem has reached such a level that, despite Official Development Assistance (ODA) flows, there is a net flow of capital from developing to developed countries in many instances. The terms of trade have continued to deteriorate, resulting in an annual loss of approximately 10 per cent of export earnings. Adedeji's comment is that:

> The cumulative toll of this unremitting decline for a whole decade on our society is clear and unmistakable. The number of countries classified as least developed among the developing world – the wretched of the earth as they have been categorized – increased from 17 in 1978 to 28 in 1988. And more, I regret to say, are knocking at the door to join. Whereas in 1960 Africa had 124 million illiterates, in 1985 the illiterate population had

increased to 162 million, almost one-fifth of all the illiterate females in the world are in Africa.

(Adedeji 1989: 11)

Equally disturbing was the situation that was revealed by the UNICEF report, *The state of the world's children*, which was released in December 1988. It showed that the impressive gains that had been made in education and health in many African countries, which appear in the ECA socioeconomic indicators (UNECA 1987), have in recent years been seriously eroded.

> For almost 900 million people, approximately one-sixth of mankind, the march of human progress has now become a retreat. In many nations, development is being thrown into reverse. And after decades of steady economic advance large areas of the world are sliding back into poverty. Throughout most of Africa and much of Latin America average incomes have fallen by 10 to 25 per cent in the 1980s . . . In the 37 poorest nations, spending per head on health has been reduced by 50 per cent and on education by 25 per cent over the last few years. In almost half of the 103 developing countries for which recent data are available, the proportion of six- to eleven-year-olds enrolled in primary school is falling . . . The slowing down of progress and the reversal of hard won gains is spreading hardship and human misery on a scale and of a severity unprecedented in the postwar era . . . For most of the countries of Africa, Latin America and the Caribbean almost every economic signal points to the fact that development has been derailed. Per capita GNP has fallen, debt repayments have risen to a quarter or more of all export earnings, share in world trade has dropped and productivity of labour has declined.

(UNICEF 1988: 1)

There is academic debate over the nature and cause of the crisis and some have even questioned the validity of the use of that term. In agriculture it is not clear, for example, if the crisis is one of production, since 'performance failures might easily be attributed to consumption, environmental or distribution crises' (Watts 1986: 377). Colin Leys (1987) argues that per capita food production in Africa may not have declined, despite World Bank (1990) estimates to the contrary, and that what we are seeing is a decline in food exports. This argument can be taken as an indicator that rural people, who are faced with a variety of problems including inadequate pricing and marketing arrangements, choose to consume agricultural surpluses rather than market them, leading to a situation where production statistics, which are usually of marketed food, go down while the per capita food production remains stable or even goes up, at least in the rural areas.

It is also evident that there are enormous regional variations in the nature and impact of the crisis both among and within African countries, but on balance there seems little doubt that rural poverty is on the increase and that the use of the word 'crisis' is fully justified. There is certainly little evidence from the case studies that, at the local level, Leys' arguments are valid.

The ECA year-end report for 1988 (Economic Commission for Africa 1990) reported a doubling of growth rate for 1988 to 2.5 per cent but this was more than offset by population growth of 3 per cent and the prognosis was that living standards would continue to decline in 1989. Adebayo Adedeji comments that: 'Even this cheerless forecast may turn out to be optimistic if the main underlying assumption with respect to (favourable) weather conditions were to turn out to be misplaced' (Adedeji 1989: 10).

THE CAUSES OF THE DEVELOPMENT CRISIS IN AFRICA

There is no simple explanation for the current state of affairs, since a complex set of external, internal and environmental factors are involved. These are, of course, closely interlinked and no one factor operates in isolation. Recently much attention in the development studies literature has been focused on the role of African states. At the 1987 UNCRD/IDS Nairobi seminar, Ngumbu Mussa-Nda argued:

> The development strategies followed by African countries during the two to three decades of their political independence has gradually led the continent into its present destitution . . . Worse still, all plans designed by those concerned with development, whether social or economic, indicated that without exception, the present policies, plans and strategies are incapable of bringing about any growth recovery in the foreseeable future.
>
> (Mussa-Nda 1988: 3)

The distinguished African geographer Akin Mabogunje later argued in a similar vien. 'It is, however, generally agreed that the false start in all African countries has been due largely to the high level of governmental and bureaucratic domination of the economy with its consequences of inefficiency, profligacy and inappropriate control' (Mabogunje 1988: 25). These are not exactly moderate statements and, as one author comments, the African state is under attack 'from left, right and centre' (Beckman 1988: 26), which has led to a serious questioning, both within Africa and without, of the role of the state in formulating and implementing policies to alleviate the current crisis. Brett sums up the situation as follows:

> State structures in the Third World are widely regarded as rigid, inefficient, overstaffed and corrupt – systematically exploiting their privileged status to provide minimal services to the public and extract monopoly rents from their clients, while conducting their business in secret to defend their activities from public scrutiny and control. So extreme is the condemnation that fundamental restructuring is usually part of the adjustment programmes introduced by the International Monetary Fund (IMF) or World Bank (WB) where the remedy is often not just the reform of existing structures but privatisation and the enforcement of control by 'market forces'.
>
> (Brett 1988: 4)

Both Brett and Beckman feel that this has gone too far, with Brett insisting 'that the role of the state must remain central in any effective development strategy' (Brett 1988: 4), and Beckman pointing towards the resilience of the post-colonial state:

> The post-colonial state is unlikely to disintegrate through some downward spiral of decay . . . Increasingly sophisticated local ruling classes are busy looking after their own houses as well as their own national development projects. In this, they are prodded, supervised, trained and financed by transnational state organs and foreign aid agencies who have their own stake.
>
> (Beckman 1988: 31)

The role of the state is obviously of critical importance to development from within and consequently will be considered in more detail later in the chapter but, before doing so, some of the external factors influencing the current development situation will be discussed.

Dame Judith Hart, quoted within Oakley 1988, draws attention to some of the more obvious linkages between external factors and the

218

ability of African states to tackle the challenges of rural poverty. She identified five areas of concern:

1 The collapse of commodity prices, leading to a substantial decrease in funds available to developing countries to finance development. She also pointed out that, given the elasticity of demand for certain commodities, the free market system is weighted against the rural poor. The ECA reported a further decline in commodity prices for 1988 (Economic Commission for Africa 1990) and the value of exports fell 2 per cent from US$ 51.4 billion in 1987 to US$ 50.3 billion in 1988. Imports were little changed at US$ 60.4 billion.

2 The adverse effects of the agricultural policies of industrialized nations which have become increasingly protectionist especially against processed agricultural goods. The EEC policies on self-sufficiency in sugar, for example, have had very negative results for sugar producers. The impasse over agricultural policies at the Montreal meetings of GATT in December 1988 has made matters even worse.

3 The combination of poor commodity prices, the debt-servicing burden, the high price of imports and the impact of the various oil stocks has resulted in a lack of financial resources to tackle poverty and a shortage of foreign exchange.

4 A decrease in real terms of net transfers of ODA. Although there was some improvement in 1988, this was more than offset by the complete drying up of new commercial loans and by the heavy weight of debt re-servicing. Adedeji comments: 'An increasing number of countries are merely deferring the debt-service problem into an uncertain future' (Adedeji 1989: 10).

5 The debt crisis that has forced governments to accept IMF structural adjustment policies which, she argued, have four main elements: a cut in wages, control of the money supply, devaluation, and a cut in public expenditure. As she pointed out:

The implication of this 'adjustment' is *nil or negative growth*. Third World countries are increasingly unable to provide the services for rural development and have also been obliged to cut subsidies for basic foodstuffs; all of which has an immediate effect on the rural poor. In Sri Lanka and Zambia, for example, the IMF *obliged* the Governments to cut subsidies which directly benefited the rural poor. In effect the IMF policies ask the rural poor to 'enjoy an even greater poverty' and have also been a direct cause of political destabilisation.

(Hart quoted in Oakley 1988: 23)

In the same month as the publication of the UNICEF study, *The state of the world's children*, the OECD released its half-yearly report (OECD 1988). The contrast between them was striking. The OECD report showed an annual growth rate in excess of 4 per cent for member nations. The leading non-Communist industrial economies were reported to be at their 'most buoyant since the early 1970s' and 'The brisk expansion has been widespread. Investment in industry has been growing especially fast' (OECD 1988: 3). The major problems which were identified were growing protectionism and the imbalance among the major industrial trading blocs. Little mention was made of the impact of the policies which have led to this resurgence of growth of the economies of developing nations yet UNICEF observes that the crisis in developing nations 'is happening not because of any one visible cause but because of an unfolding economic drama in which the industrialized nations play a leading role' (UNICEF 1988: 1).

It is important when considering the context for development from below in Africa to examine the relationships which exist at various scales as well as the major actors involved and the relationships among them.

Regardless of which actors are involved in the struggle for development in Africa or at which scale the development problematique is considered, the influence of environmental factors looms large. By far the most significant factor is rainfall. The dynamics of Africa's climatic patterns are complex and by no means completely understood. There is also a tremendous variety and variability of climatic factors which makes the concept of mean annual rainfall almost meaningless in the African context. Three facets of rainfall must be considered: seasonality, reliability and rainfall type. Seasonality is perhaps the best known of these facets. Much of intertropical Africa has a single rainy season although areas closer to the equator have two. In the latter case, the second or 'short rains' period is much more variable and much less reliable than the first or 'long rains' period. Variability and lack of reliability are a feature of Africa's rainfall regimes. Even in areas of high mean annual rainfall, the possibility of drought in at least one year in ten is high. As mean annual rainfall totals decrease, the incidence and impact of droughts become even greater. The droughts of the 1980s, which are mentioned in several chapters of this book, are by no means unique. Their dramatic economic and social impact puts great strains on the coping mechanisms of African societies and drought has played a major role in causing the crisis that is facing many of Africa's peoples. Variability can range up to 100 per cent of the mean for many regions of Africa, which signifies that a recurring problem is not only drought but also too much rainfall in some years.

Much of the rain tends also to fall in very intense showers, with 100–200 millimetres of rain in a matter of a few hours being not at all unusual. For agricultural purposes this type of rainfall poses problems of rapid run-off and soil erosion.

Most of the soils of intertropical Africa are not very fertile, with a high percentage of lateritic soils which rapidly lose their fertility when cleared of indigenous vegetation.

The African environment is difficult for both agricultural and pastoral peoples and, as it comes under increasing population and production pressure, the difficulties increase.

The impact on agricultural production is direct but what is often forgotten is that the availability of water is critical for all economic activity. Many industries use water more intensively than agriculture. The nature of Africa's physiography and topography makes both access to ground water sources and the use of irrigation difficult and expensive, and so the economic dependence on rainfall becomes even greater.

Both the external and internal factors that have been discussed as causes of the crises facing Africa are probably of less consequence than the impact of Africa's weather and climatic patterns over the last decade.

THE INADEQUACIES OF PAST THEORY AND PRACTICE

Robert Chambers (1989b) argues that 'Historically, the fashions for ideologies, packages and programmes in rural development have changed' and he claims that this is a reflection of changing conditions:

> The lesson is to see ideology and action in context, not as constants, but as arising from and adapting to, as well as moulding, these conditions. In this view they are always likely to be out-of-date, always requiring an imaginative effort to be ahead of current convention.
>
> (Chambers 1989b: 1–2)

221

He discusses in detail the inadequacies of both the neo-Fabian approaches of the 1970s, which are categorized through redistribution with growth, and the neo-liberal approaches of the 1980s which emphasize structural adjustment and market forces.

A major inadequacy of both approaches has been that they have given inadequate attention to the realities at the local level.

> Both ideologies, and both sets of prescriptions, embody a planner's core, centre-outwards, top-down view of rural development. They start with economics, not people; with the macro not the micro; with the view from the office not the view from the field. And, in consequence their prescriptions tend to be uniform, standard and for universal application.
>
> (Chambers 1989b: 6)

It could also be added that both approaches have so far failed to bring about the results that were expected of them and that African governments which have pursued them have met with little success. It can be argued that any macro-level solution to development in Africa, regardless of its theoretical or ideological perspective, has severe limitations and nowhere are these limitations more obvious than at the micro scale of the African local community, as the case studies in this book amply illustrate. Chambers argues for 'a counter-ideology of reversals' where development theory and practice are driven by 'putting first the priorities of those who are few and peripheral' (Chambers 1989b: 9). Both development 'from below' as originally conceptualized and development 'from within' are such strategies. The general solutions and explanations that are so common in development theory and practice must be complemented by not only an awareness of the need for diversity and specificity but also a recognition that the diversity is critical. 'For some professionals, development is still, consciously or unconsciously, seen as convergent; in the paradigm of reversals, development is decentralized and divergent' (Chambers 1989b: 19).

It is only 'by becoming more complex and diverse that ecosystems and livelihood strategies become more stable and more sustainable. Near the core of this paradigm is decentralised process and choice' (Chambers 1989b: 20). If meaningful development is to occur, it must be defined, motivated and controlled to a much greater degree than at present by the local population itself. The rationale for local initiatives is so compelling that the two major questions are: how can effectiveness be improved and how can they be made more central to development theory and practice? Following what Chambers (1988) calls a complete reversal of normal bureaucratic procedures, development needs to be turned on its head and to be viewed in a radical way. The word 'radical' is used here in the original meaning of the word, namely, to look at development from its roots. To use a biological analogy which is appropriate to Africa, that part of a maize plant which we do not see, the root system, is much more extensive and complex than that which we do see. Local initiatives and participation have become the imperative of the development decade of the 1990s.

THE CASE FOR LOCAL INITIATIVES

In February 1989 an international seminar was held in Arusha, Tanzania in which four of the authors of this book, Zinyama, Ondiege, Ndaro and Taylor, participated. The theme of the seminar was 'Reviving Local Self-Reliance: Challenges for Rural/Regional Development in Eastern and Southern Africa' and it was jointly sponsored by the United Nations Centre for Regional Development (UNCRD) and the Centre on Integrated Rural Development for Africa (CIRDAFRICA). Many of the ideas in this chapter are built on the opening paper of that seminar, 'Why local initiatives in Africa?: the context and rationale' (Taylor 1989). Seven major arguments were advanced about the necessity for local initiatives and these can be summarized as follows.

1 In the current state of crisis in Africa, local-level initiatives are necessary for survival.
2 African governments, either by choice or by necessity, are unable to provide adequate development services and local people cannot depend on the state to provide them with sufficient means to improve their quality of life.
3 There are unutilized or underutilized resources available at the local scale, including financial resources, which could be mobilized for development purposes.
4 The enormous diversity in rural Africa in economic, ecological, sociological and political terms is such that development can be effectively addressed only at the local level. Chambers captures this well:

Local diversity has many social and ecological dimensions, both within and between areas. Social diversity has many aspects – ethnic, cultural, economic (concerning wealth, poverty and

223

access to and control over resources), educational, occupational, gender and age among others . . . Then there are many forms of ecological diversity . . . Differences of soil, slope, vegetation, multiple canopies of plants, multiple tree–crop–livestock interactions, and the number of species exploited, can be mind-blowing. And finally there is diversity which is regularly seasonal and irregular in interannual variation. Nor is this all. Social and ecological diversities interlock and multiply variance. It is easy, once one starts seeing and thinking this way, to regard each place and social group as unique, requiring its own path for development.

(Chambers 1988: 51–2)

5 Economically sound 'sustainable development', as emphasized by the Brundtland Report (World Commission on Environment and Development 1987), has unique meaning at the local level. In terms of the physical environment, it is at the local scale in Africa where environmental degradation or preservation is most likely to occur. Environmental degradation by Africa's rural people is not an act which is bred out of ignorance and greed, as suggested by some authorities, but more often is a result of desperation and the need to survive. Local people have a very special interest in the sustainability of any development initiative. External actors often define policies and impose these by a variety of means on local people. It is the local people who have to live with the results and their caution in adopting new policies is often more than justified.

6 Central planning has inherent weaknesses and the need for decentralization and devolution is paramount not only in Africa but in many other countries, both 'developed' and 'developing' and both capitalist and socialist.

7 Indigenously based knowledge has a major contribution to make to the development process and this local knowledge is effectively found at the local scale.

The importance of local initiatives should not be romanticized. As Mackenzie argues in Chapter 1, the situation is highly complex and a central question is whether grass-roots initiatives are truly strategic in intent and lead to sustainable development action or are merely survival strategies to cope with crisis. Both neo-Fabians and neo-liberals have adopted with enthusiasm the rhetoric of local participation but primarily as an extension and modification of the central premises underlying both paradigms. The answer to some of these questions lies in the evidence from detailed case studies, such as those in this volume, and it is to them that this chapter now turns.

DEVELOPMENT AND SURVIVAL: THE EMPIRICAL EVIDENCE

Chambers has argued that, in order to understand local realities, it is necessary

> Through local study and individual cases . . . [to] show how varied is that universe of vulnerability and poverty for which we seek simple explanations and single solutions . . . the lesson for the future is to enquire and question, doubting what we think we know and learning from and with those who are vulnerable and poor . . . and to do this, not in one locality, and not for one group only but again and again, in each place and for each sort of person. For that is the surest path to better understanding and to action that will better fit and serve the diversity of conditions and people and their changing priorities and needs.
>
> (Chambers 1989a: 7)

The eight case studies in this book contribute to the empirical base that is required to improve our understanding at the local level.

The case study by Zinyama of local farmer organizations in Zimbabwe is interesting. Zinyama argues that, in Zimbabwe, the government is not retreating but is supportive of local initiatives and that it is therefore not appropriate to consider such initiatives as reactions to the withdrawal of the State from rural areas. Local-level participation has been facilitated by the democratization of local government, a point which is also emphasized by Olowu (1989). Zimbabwe has experienced remarkable production increases in the small-farmer sector since independence in 1980, but these mask both spatial inequalities and intra-community differences of age, gender and status.

> The benefits from the new rural development thrust are largely accruing to the small number of peasant farmers who are fortunate enough to be located within the better agro-ecological regions. Elsewhere, farmers continue to be handicapped by the constant threat of drought and food shortages, by increasing land shortages, by infertile soils and by low agricultural production.
>
> (Zinyama, Ch. 2: 43)

He reminds us that the variability of the African physical environment is a major factor and that the relative impact of any policy or institutional intervention depends very much on the agro-ecological conditions of the area concerned.

Zinyama looks at the history of various forms of local organizations in two communal areas south of the capital city of Harare, including mutual-help groups, farmer training groups, agricultural marketing groups and voluntary savings or thrift clubs. These have achieved considerable success, which Zinyama argues is due to two major factors: the complementarity between the efforts of the State and those of the rural population itself; and the fact that, especially for the Master Farmer clubs, the groups have voluntary membership, self-government and peer control. Voluntary savings clubs have the additional advantage that, in the event of crop failure, farmers purchase inputs out of their own savings rather than through loans.

The evidence from this case study suggests that in Zimbabwe grass-roots initiatives are truly 'strategic in intent' (MacKenzie, Chapter 1) and are much more than simply a means of coping with crisis.

The three case studies from Ghana are quite different from each other and illustrate the diversity at the local level that was outlined earlier in this chapter. George Dei, using participant observation techniques, outlines the adaptive responses of the people of Ayirebi in South-eastern Ghana not only to local food supply cycles but also to a national economic crisis of the 1980s which was triggered by world recession and was aggravated by drought, bush fires and an unexpected influx of Ghanaian deportees from Nigeria which led to a 7 per cent increase of population in the town.

The community managed to cope in a truly remarkable way in the circumstances and Dei's evidence suggests that, despite community differentiation, the community reacted as a group to the crisis facing them. He talks of 'a remarkable degree of co-operation between genders' (Dei, Ch. 3: 73) and of communal cultivation of new land in the form of co-operative farms. Richer farmers who had food surpluses were persuaded by community pressure to offer these at reasonable prices to the poorer members of the Ayirebi community first before attempting to sell elsewhere.

The local explanation for the crisis is also interesting. The people are reported to have seen this as owing 'to the breakdown of respect for customs, including one's obligation to kin and neighbours' and to have explained the drought as 'punishment from the gods and ancestors in order to make the living aware of their neglect of adaptive customary behaviour and of the need to live harmoniously in the community' (Dei, Ch. 3: 73–4).

Residents of Ayirebi who had migrated to the cities and other regions of Ghana also increased their remittances in both cash and kind to help the community in its time of need.

This case study lends support to the continuance of the 'organic African community' which appears to have been strengthened rather than weakened by crisis. Mackenzie argues in Chapter 1 that it is a mistake to 'mythologize' undifferentiated local communities where the interests of all 'are served through subscription to community development efforts' (Mackenzie, Ch. 1: 27). Although Ayirebi is not an undifferentiated community it appears to have responded to crisis as a community. Whether this is a coping mechanism in the face of crisis which is unlikely to be maintained or a continuing strategy which can lead to sustainable development remains to be seen. The community appears to be meeting its basic needs and has taken steps to increase equity and redistribution. It also appears to have taken cognizance of the qualitative transcendental values which it was argued earlier are an element in defining development. Ayirebi does not appear to be challenging the interests of those in power at the State level but is in fact using the organizational mechanisms of the State, such as the Committee for the Defence of the Revolution and the Town Development Committee, for its own ends. This is a political act but of a different kind from those which are usually thought of as leading to local empowerment. The Ayirebi case does contain the 'latent seeds for multiple new developmental actions' (Goulet 1989: 167) but these would appear to be limited in scope. Dei calls for the State to revise its views on an externally focused, export-led strategy and to build 'thriving, self-reliant, self-sustainable local and regional communities in contemporary Africa' (Dei Ch. 3: 75).

The other two case studies from Ghana lend no support to the existence of an undifferentiated 'organic' community and show evidence of considerable tensions and conflict at the community level.

Songsore analyses the history of the co-operative credit union movement in North-West Ghana, an area which he describes as 'the worst pocket of extreme poverty in Ghana' (Songsore, Ch. 4: 84). Although the co-operative credit movement is an innovation which was introduced by a Canadian Catholic priest, it builds on traditional rotating credit institutions known as 'susu'. The formal banking system has little interest in this poor region which has been hit very hard by the negative impact of the Structural Adjustment Programme that was introduced at the urging of the World Bank. As the area is ecologically unsuited to the growing of cocoa, it has benefited little from the export-led development strategy which has concentrated resources on the cocoa growing areas of southern Ghana. The inter-regional barter terms of trade have shifted against the rural people of this region and the introduction of user charges has hit social pro-grammes, such as health and education, very hard. Women and children in particular have suffered. To the disadvantaged people of this disadvantaged region, the suggestion that the Structural Adjust-ment Programme is 'succeeding' in Ghana is a travesty of the truth. It is making an already difficult situation much worse.

In these circumstances, credit could be an important tool and the aim of the Co-operative Credit Movement was 'to serve as a focal point for the endogenous, grass-roots mobilization of local resources for local development within the limits set by the national and inter-national system' (Songsore, Ch. 4: 85). Songsore's analysis shows how an institution which was originally established to protect the poor was gradually taken over by members of the local dominant classes. 'Credit unions, by supporting investments in the productive base of the regional economy, have been vital instruments in rural develop-ment' but when, in the 1970s, the Catholic Church withdrew its priests as treasurers the replacement treasurers 'soon became pawns in the hands of local notables' (Songsore, Ch. 4: 94). The result was a crisis in confidence in the co-operative credit movement and an increase in corruption and misuse of funds. Active membership dropped from 25,830 in 1983/4 to 16,290 in 1986 and loan defaults increased in scale. The major defaulters were the 'big men'. 'Poor peasants, rural women and workers are rarely mentioned, if at all, among the list of loan defaulters' (Songsore, Ch. 4: 95) Songsore argues that the Co-operative Credit Union can be of real value but it must be reformed: 'recapitalizing and redirecting the credit unions towards the genuine developmental aspirations of the rank and file membership, especially women' (Songsore, Ch. 4: 85).

This has already begun by a combination of both local and outside action. The Canadian Co-operative Association and the Co-operative Credit Unions Association of Ghana have formed a new alliance to help to revive peasant control of the co-operative credit union in North-West Ghana and to recapitalize the movement. In some areas the peasants have taken action in their own hands.

In Nandom the situation was so grave that peasants in a fit of anger and protest literally drove off the treasurer and some other committee members . . . From the mess has emerged a farmers' co-operative credit union, the Kuob-Lantaa Credit Union. The peasants have refused to admit any literates into the new union, since the literate 'elite' were blamed for the frauds in the Nandom Credit Union.

(Songsore, Ch. 4: 97)

Development from within in North-West Ghana will have to come through strategies of revival and peasant control which are based on women, who are in the majority in the region. Such strategies may not be possible without outside help and support from non-government organizations.

Takyiwaa Manuh's study of the salt co-operatives of Ada District lends support to the argument made by Mackenzie that 'To call for the "empowerment" of local people is to challenge social structure. Profoundly, one is dealing with "politics" not "policies", with "struggle" and not "strategy" ' (Mackenzie, Ch. 1: 1). Manuh argues that 'it is clear that, for most rural dwellers, survival and the little development that has come their way have largely resulted from their own efforts' and that the salt co-operatives are 'an instrument of struggle and a coping strategy of rural people, in the face of dispossession by powerful capitalist interests and state neglect' (Manuh, Ch. 5: 103).

In the 1970s, two private companies were given leases to win salt from the lagoon, thus replacing traditional groups. The area of the lagoon that was left for the Ada Traditional Council was not suitable for both ecological and religious reasons. The locals formed the Ada Songor Salt Miners' Co-operative Society to act as a focus in their struggle to regain their rights. It is interesting that they utilized the form and regulations of other co-operatives in Ghana and that they operate under licence from the Department of Co-operatives like other societies but that they are quite different in both form and function.

The main protagonists are Vacuum Salts Limited, which is owned and managed by a powerful Ghanaian family, the Appentengs, and the local peasant communities living around the lagoon. The Central State has been equivocal in this struggle, orginally allowing a takeover of Vacuum Salts and then restoring it to the Appenteng family. When the Co-operative began to show success, the State was quick to impose new taxes on their salt production both centrally and through its local agents, the District Council. The Ada Traditional Council, which is controlled by the Chiefs, appears to have allied itself with the District Council and is also extracting levies. The main spokesperson for the local community, the president of the Ada Songor Salt Miners' Co-operative Society who is also a State Assemblyman, was dismissed from office by government in July 1989.

It would seem that local people face repression when they become too vocal in their demands for control of their own destiny and confront the State and its allies. It is likely that local initiatives in such circumstances will face sustained conflict. It is also clear that traditional structures, such as the Ada Traditional Council, do not always act in what local communities perceive to be their best interests. Communities such as Ayirebi appear to be 'invisible' to the State and do not appear to threaten its interests. The communities around the Songor Lagoon are a very different matter.

The case study of Machakos District in Kenya by Ondiege is very much a planner's perspective of local initiatives. He argues for 'an enabling environment that will allow the participation of local rural people in conceptualizing, implementing and managing their development programmes' (Ondiege, Ch. 6: 127). In Kenya the State is clearly trying to incorporate local initiatives into its planning strategy and local initiatives are seen as an extension of this strategy. There is no doubt that mobilization is taking place, with 96 Self-help and 249 Women's groups in the District together with a number of Co-operative Societies, but it is not clear if these are a reflection of the people's perspectives of development priorities or those of the government.

It is possible that there is a convergence of these two perspectives especially in the field of income-generating activities. The budget figures show a high degree of dependence on external funding. In 1986–7, out of a budget of 3.8 million Kenya Shillings for these activities, a full 3.6 million came from foreign aid agencies and non-government organizations with only 155,387 shillings from central government and 64,370 shillings from Machakos District Council. This lends credence to the view that this form of local development is the 'cheap "development platform" (to paraphrase Watts, 1989: 6)' referred to by Mackenzie at the beginning of Chapter 1. On the other hand, it can be argued that it is in the best interests of the people of Machakos to participate in these local initiatives, since, regardless of who initiates them, the benefits accrue to the participants.

The second Kenya case study by Noel Chavangi describes a special programme of the Ministry of Energy to persuade farmers to plant fuelwood, which is a critical input for development in rural Kenya. The Kenya Woodfuel Development Programme is supported by the Swedish International Development Agency. Again, the initiative is clearly an extension of government development policy but an innovative attempt is being made to involve local communities in project identification and design and to incorporate indigenous expertise and practice which is acceptable and appropriate to each cultural setting. The objective is to increase woodfuel production both for self-consumption and for the market and here it appears that the government, in developing its programmes, is making a genuine attempt to listen regarding the realities facing the farmers of Kakamega District. The project is, for example, making a special effort to involve women, who will be major beneficiaries, and is using techniques such as popular theatre to expose issues such as gender relationships which are involved. The main barriers to increased fuel-wood production are not lack of knowledge or technical in nature but are sociocultural. Here there has been at least some of the 'bureau-cratic reversal' that was called for by Chambers (1988), but the case study is one of mobilization by government rather than local initiatives from the people. The pragmatic view would be that if it increases the supply of much needed fuelwood, which meets local needs, then the form and context of participation are of less concern. If government perceptions of development priorities and those of local people coincide then conflict is unlikely. However, there is an interesting contrast between the government's use of popular theatre in the case of the Kenya Woodfuel Development Programme and its repression of the Kamiriithu Community Education and Cultural Centre, which used the same technique to attack government (referred to by Mackenzie, Ch. 1: 29–30). Once again, if development from within confronts the State in a direct challenge, it is likely to be repressed.

Japheth Ndaro, in discussing local coping strategies in Dodoma District in Tanzania, also takes the view that local initiatives which are not part of the government's plans are likely to remain marginal to the development process. He argues that there have been three phases of local coping strategies since Tanzania attained independence. In the first phase, which he suggests concluded in 1970, the

231

people in Dodoma District devised local coping strategies which did not fit the government's plans. These involved trade, crafts, industries, the building of small dams and the bringing of new land under cultivation, as opposed to schools, dispensaries and roads which is what the government wanted. Ndaro sees these as 'survival' as opposed to 'development' strategies but it could equally well be argued that the people's preferences did not fit the top-down model of development being promulgated by the government. In the second phase, from 1971 to 1978, villagization was the key factor. During the third phase, which is post-1979, Ndaro argues that the government has stifled rather than encouraged local initiatives. The range of strategies that is described for Dodoma is impressive and people seem to be finding ways to survive in difficult circumstances but at issue is whether these can be sustained over time and how meaningful they are in the current Tanzanian context.

Alice Nkhoma-Wamunza's case study of the women brewers of Utengule Usangu village in the Mbeya Region of Tanzania is an example of how women organized to improve their economic welfare and to retain a greater degree of control of the fruits of their own labours. She illustrates the struggle that women have in a male-dominated society and how, in the long run, their initiatives were taken over by a male-dominated 'community' in the name of the welfare of the wider community and to the detriment of the women of that community. A promising development from within initiative was stifled not because the women threatened state power but because they confronted male hegemony at the village level.

The empirical evidence that is presented in these case studies is both rich and valuable and complements a number of other similar studies by African authors, which will be used to consider the central theme of this chapter which is to point to new directions for development theory and practice. These include studies on Tanzania by Mbilinyi (1989), Kapinga (1989), Ndara and Temu (1989), and Maro (1990); studies on Kenya by Mbugua (1989), Kobiah (1985) and Imoo and Louse (1988); studies on Zambia by Chileshe (1989) and Kalapula (1989); studies on Zimbabwe by Macebo (1988) and Zinyama (1989); a study on Uganda by Ouma (1989); and a study on Ethiopia by Asgele (1989).

DEVELOPMENT FROM WITHIN: KEY COMPONENTS AND PROSPECTS FOR SUCCESS

Development from below revisited

Development 'from within' is based on the concepts of 'development from below', as outlined in *Development from above or below?: the dialectics of regional planning in developing countries* (Stöhr and Taylor 1981). It is perhaps useful to revisit some of the ideas of 'development from below' before considering 'development from within'. Although new in the context of its time, the concept had its roots in the populist ideas of the nineteenth and early twentieth centuries. A number of books advocating broadly similar or related issues appeared about the same time. These included: *Territory and function: the evolution of regional planning* (Friedmann and Weaver 1979), *Self-reliance, a strategy for development* (Galtung *et al.* 1980) and *Alternative Raumpolitik* (Naschold 1978) in German.

Development from below was primarily developed in a 'Third World' context and grew out of a synthesis of a number of different ideas which were broadly related to the emerging number of 'alternative development' strategies. It was influenced by a re-examination of populist and anarchist thought of the nineteenth century, allied to the major contribution of thinkers such as Julius Nyerere and Mahatma Gandhi. Development from below was also strongly influenced by dependency theory and by the concept of an ecologically sound development as advocated by Sacks and his colleagues. Shumacher's concept of 'small is beautiful' and appropriate technology also played a part. The concept of development from below saw development as an essentially indigenous process in which concepts of self-reliance and popular participation loom large. Development from below was based on the maximum mobilization of each area's natural, human and institutional resources, with the primary objective being the satisfaction of the needs of the inhabitants of that area. The dominant building block was the rural, territorially based community at the smallest scale that is efficient and effective. The strategy was basic needs oriented, labour intensive, ecologically sensitive, regional resource based, rural centred and argued for the use of appropriate rather than highest technology.

The concept has now been adopted in rhetorical terms and the slogan 'development from below' has entered the jargon of regional and development planners at all levels. However, there is a lack of specificity of what constitutes 'development from below' and a wide

233

variety of interpretations is given by those involved in dealing with it. Some of these bear little resemblance to the original paradigm and many are in fact antithetical to the original concept.

The original concept was criticized from a number of perspectives. It was argued that there were three major shortcomings: inadequate specification of the theoretical underpinnings of development from below; failure to specify the necessary and sufficient conditions in which development from below could emerge; and failure to add an adequate theory of explanation to what in essence was a theory of policy.

There was also another strand of criticism which did not appear in the literature but was nevertheless very real. Considerable scepticism was expressed by some indigenous planners who said that development from below and concomitant ideas, such as agropolitan development, were just one more example of theories and prescriptions which are developed in the North being applied to the South. It was argued that development from below, before it could possibly be taken seriously, would have to be applied to the industrial nations of the North. There was a suspicion that what was being suggested was a palliative, on acceptance of the inequities of the international system, rather than a device to achieve meaningful change.

Strangely enough, the subsequent development of the concept has taken place almost exclusively in a European context. In the 1980s, several studies on the topic were published, including: *Self-reliant development in Europe* by Bassand *et al.* (1986), *Regional analysis and the new international division of labour*, Moulaert and Salinas (1983), *Economic restructuring and the territorial community*, Muegge *et al*, (1987), and *Endogenous development*, Stuckey (1985).

Self-reliant development in Europe is perhaps the most comprehensive of these and reveals some current thinking in the field. The book itself indicates that interest seems to have shifted from the problems that are inherent in poverty to a means of dealing with the malaise of post-industrial society. A quotation from the Introduction by Brugger and Stuckey will illustrate this:

> The demands of economic competitiveness and economic growth conflict more and more frequently with a growing concern for self-development, social morality, and territorial and ecological integrity. We are witnessing a new longing, a longing which reveals a shift in values: from functional goals to territorial life-space, from material gains to emotional and

spiritual needs, from one-sided intellectual training to meet the demands of the computer age to questions about human life and the natural environment. The revolt of youth, the peace movement, concern for health, natural foods and alternative medicine, the continued drive for ecological sanity – despite rising unemployment – are all examples of a new interest in the concept of self-reliance or development from below.

(Brugger and Stuckey 1986: 1)

New terms such as 'endogenous development' have been coined and new meanings given to others. John Friedmann, for example, defines self-reliance as a form of radical social praxis. He argues: 'A self-reliant society is an inclusive, non-hierarchical society that stresses co-operation over competition, harmony with nature over exploitation, and social needs over unlimited personal desire. It represents the one best chance for the survival of the human race' (Friedmann 1986: 211). In yet another apparent shift of thinking, Friedmann, who originally proposed endogenous development as a solution for poverty and dependency in peripheral regions, now argues

that the route to endogenous development within the mainstream of economic policy is virtually closed. It is a viable option only for world city regions that can use their countervailing power to negotiate with global capital and with the state for arrangements favourable to themselves, or to be more precise, to their political and economic elites.

(Friedmann 1986: 211)

He argues further that 'it is virtually impossible to create regional enclaves modelled on social relations that are essentially different from those of the system in dominance' and that hopes to the contrary 'are always dashed [on] the bedrock of reality' (Friedmann 1986: 212). Given this, he therefore talks of 'self-reliant' development as distinct from endogenous development but he sees self-reliance as 'a strategy of social mobilisation for political ends' (Friedmann 1986: 213). He talks of militants organizing people to confront specific questions in the public sphere and to change reality through their own actions. He argues that these militants should use the peripheral regions and the 'internal peripheries' of cities as 'staging areas' for their self-reliant praxis, which he sees as the only viable option that peripheral populations have for their survival and well-being. He sees his self-reliant approach as very much development from below and he claims to have derived it from a combination of utopian, anarchist and marxist thought. Self-reliant development is a way of becoming politically engaged and its ultimate strength 'derives from a deliberate disavowal of formal organisation. The object of the movement is to change reality not administer it' (Friedmann 1986: 213).

The book contains a variety of interesting studies from Switzerland, Austria, France, Belgium, Italy and Scotland, together with three chapters on the historical and theoretical context. All of the authors are purported to share a common conviction in development from below but it is clear that there is little agreement among them on many issues. Even an informed reader who is sympathetic to the paradigm may find the confusion and contradictions difficult to deal with. The relevance of some of the ideas to the developing nations of the world is open to question.

Development from within has similarities with the original concept of development from below but little congruence with current usage of the term in the literature cited above. Development from within is not only possible but also offers the best chance for survival in Africa's current crisis and the key components of development from within will now be discussed.

Components of development from within

Participation

Participation is a key component of development from within. It is a concept about which much has been written (Oakley and Marsden 1984) and a recent article by Goulet (1989) gives a good overview of some of the main issues. Following Wolfe, participation is defined as 'the organised efforts to increase control over resources and groups and movements hitherto excluded from such control' (Wolfe quoted by Goulet 1989: 165). Goulet argues that there are many kinds of participation and suggests a fourfold typology classifying participation in terms of: (a) participation as a goal or as a means; (b) the scope of the arena in which participation operates; (c) the originating agent of the participation; (d) the moment at which participation is introduced.

This typology provides a useful framework for consideration of the role of participation as envisaged in development from within. Participation is seen as both a goal and a means; it operates primarily at the local community level in the first instance. It is not induced

236

from above but is generated from below by the populace itself; it can also be generated by the catalytic action of some external third agent. In terms of the timing of the involvement, it begins with the first step of Goulet's sequence: the initial diagnosis of the situation.

Goulet describes the debate between authors such as Freire, who sees participation primarily as a goal and 'some problem solvers [who] defend popular consultation on the grounds that it is the best way of getting the job done or achieving lasting results' (Goulet 1989: 166). Development from within is both a theoretical and a practical concept and as such both ends and means are important.

The scale at which participation operates is at the level of the local community. This is not to suggest, as was the case with the original conception of development from below, that the whole community must always participate as an organic entity. Development from within recognizes the importance of community but also acknowledges that community may sometimes be too large an aggregation and that within communities there is considerable differentiation which is based on issues such as class and gender. This was certainly apparent from the evidence from the case studies. For development from within, participation in all cases will be at a scale much lower than that of the nation state on which much current development thinking is based. It will also be larger than the household. As Goulet has pointed out, 'Depending on the scope of the arena or field in which participation occurs, its impact on development will vary accordingly' (Goulet 1989: 166). In the first instance, this may well lead to the creation of 'islets of social organisation, which obey their own rules of problem solving irrespective of dominant rules governing society at large' (Goulet 1989: 168). An example from the case studies would be the community of Ayirebi, which is described by Dei in Chapter 3. Such an initiative will probably be a necessary first step to ensure that basic needs are met and it has value in its own right at a particular time. In the longer term, however, it must develop further so that local people 'master larger issues transcending the boundaries of their immediate problems' (Goulet 1989: 176). There must be a strategic impact at a larger scale. The best example from the case studies is the National Farmers' Association of Zimbabwe (NFAZ) to which the Master Farmers' Clubs, described by Zinyama (Ch. 2: 51ff.), belong. As Bratton (1990) has pointed out, the Association had in 1988 4,500 clubs nationwide and a membership of 70,000 farmers plus 150,000 other participants and 'can claim without exaggeration to be the only independent, national small-farmers' union in sub-

237

Saharan Africa' (Bratton 1990: 101). Control of each club remains firmly at the local level but NFAZ plays a significant policy role at the national level and Bratton gives several examples of the importance of this. The aim of NFAZ is to ensure that the problems and views of small farmers are made known to the government and it appears to do this effectively and has considerable political impact.

Participation can come from three quite different sources: 'it can be induced from above by some authority or expert, generated from below by the non-expert populace itself, or catalytically promoted by some external third agent' (Goulet 1989: 166). Development from within does not accept Goulet's use of the term 'non-expert' to describe the local populace. The main agent in inducing participation from above is government and some authorities prefer the term 'mobilization' to describe this process. Governments often view participation as a method of achieving their own goals. Development from within is rarely if ever achieved by a mobilization approach from above except possibly in circumstances where the views and needs of a local community and the goals of the government coincide. But, even in these circumstances, if the local people do not have the decision-making power, true development from within cannot occur. 'Although such "participation" is easy to promote, only with great difficulty can it achieve authenticity. Authenticity means locating true decisional power in non-elite people, and freeing them from manipulation and co-optation' (Goulet 1989: 168).

Bottom-up participation is a key to development from within and the empirical evidence from the case studies and the additional studies cited shows that in Africa today this is largely the result of the 'deliberate initiatives taken by members of a "community of need" ' (Goulet 1989: 167).

A third source from which participation can originate is what Goulet calls an 'external third agent'. In the case studies, the role of external agents such as non-government organizations has been shown to be of importance. Although generated from outside the local community,

> intervention by third party change agents differs in important respects from top-down participation induced by the state or other elite groups. Like the form initiated from below, third party participation usually aims at empowering hitherto power- less people to make demands for goods, not to contribute their resources to someone else's purposes.
>
> (Goulet 1989: 167)

There are occasions when development from within can be facilitated by outside assistance, especially in cases such as North-West Ghana, as described by Songsore in Chapter 4, where the community has been progressively weakened over time. This carries with it inherent problems, however, especially if the change agent has an agenda of its own, which is often the case with many religiously motivated institutions. The case study of Turkana by Imoo and Louse (1988) is an apparent example of this.

The Catholic Church initiated a Diocesan Development Education Programme in Turkana District in 1983 which generated a number of promising projects. 'This was the first time that the Turkana people had thought of, discussed and implemented a project on their own, without the direct assistance of a priest, nun or government worker' (Imoo and Louse 1988: 83). As the project progressed, however, problems arose.

> The emergence, therefore, of local indigenous leaders from the Adult Education Programme became a threat to the power and authority of the Catholic Church in Turkana in general and of the parish priest in particular . . . The Church realised that a conscientisation programme like the Adult Education Programme could not be used to divide people into 'Catholics', 'Muslims' or 'Protestants' but to encourage and promote a unity which could eventually become problematic to the institutional Churches. They realised the programme was producing independent thinkers. The Church realised that it had started a programme which was not a vehicle to evangelise people, which was the work all the missionaries had come to do.
>
> (Imoo and Louse 1988: 86–7)

Despite outside evaluations suggesting that the programme should continue 'because it was helping the Turkana people become participants in their own development and decision makers in their historical development' (Imoo and Louse 1988: 87), the Church withdrew its support and closed the programme in February of 1986.

The final point in Goulet's fourfold typology relates to the moment at which participation is introduced. Goulet argues that there is a discernible sequence, beginning with the initial diagnosis of the situation and leading up to discernible final action.

> At any point in the sequence, a non-expert populace may 'enter in' and begin to share in its dynamics . . . Therefore, if one

wishes to judge whether participation is authentic empowerment of the masses or merely a manipulation of them, it matters
greatly when, in the overall sequence of steps, the participation
begins.

(Goulet 1989: 167)

Development from within depends upon authentic participation and
the local populace must enter the participation process at the first
stage and be in control of all subsequent stages of action.

Authentic participation is a necessary condition for development
from within, although it is not sufficient in itself.

Participation, or some active role playing by intended beneficiaries, is an indispensable factor of all forms of development
. . . it is the nature and quality of participation . . . which
largely determine the quality of a nation's development pattern
. . . a policy bias in favour of authentic participation correlates
highly with genuine development.

(Goulet 1989: 175)

Goulet's article concludes with the following paragraph:

Participation began largely as a defense mechanism against the
destruction wrought by elite problem solvers in the name of
progress or development. From there it has evolved into a preferred form of 'do-it-yourself' problem solving in small-scale
operations. Now, however, many parties to participation seek
entry into larger, more macro, arenas of decision making.
Alternative development strategies centring on goals of equity,
job creation, the multiplication of autonomous capacities, and
respect for cultural diversity – all of these require significant
participation in macro arenas. Without it, development
strategies will be simultaneously undemocratic and ineffectual.
Without the developmental participation of non-elites, even
political democracy will be largely a sham.

(Goulet 1989: 176)

From the case studies in this volume and the additional studies cited it
is clear that in most cases African communities are at the 'defence
mechanism' or 'do-it-yourself' stage. There is some evidence of
participation reaching into the macro arena, such as in Zimbabwe,
and of potential for this to occur, such as with the co-operative credit
unions in Ghana, but as yet this evidence is limited. This is not surprising, given both the colonial heritage and the strongly centralized

state planning approaches which superseded them. However, the upsurge of initiatives that is recorded in the empirical evidence is a promising sign of new directions.

Territoriality

A second major component of development from within is that it is a territorial concept. This is seen as being quite different from a spatial concept. Gore, in his stimulating book *Regions in question* (1984), argues that regional development 'theory' has been plagued by what he calls the incomplete relational concept of space. Territory is defined here to include place and the social relations and power inter-actions which take place within that bounded space. Place has real meaning to most African peoples and it goes well beyond limited economic concepts such as ownership of the means of production. Attachment to place remains, despite physical separation over both time and space. Most Africans, even if they have moved elsewhere, wish to be buried in their community of birth and if possible on their ancestral land. A recent court case in Kenya in 1987 (Wambui Otieno v. Ougo and Sinanga), concerning where a criminal lawyer (S.M. Otieno) was to be buried, is an interesting example of how powerful is the attachment to place of birth. It also helps to explain why the body of David Livingstone was carried by his African followers hundreds of kilometres to the coast after his death.

The attachment to place is of course an extension of the attachment to land, the significance of which is captured well by Wolde-Mariam:

> The very humanity of the person and his status in society is defined mainly by his ownership of land, or more correctly by his membership to a landowning kinship group. Possession of land is therefore sought not only for economic benefits. In fact this is of secondary importance. It is sought more for the status, respectability, dignity and pride that go with ownership of land.
> (Wolde-Mariam quoted in Odingo 1988: 27–8)

The quotation, although indicative of the importance of attachment to both place and land, raises the issue of gender relationships. Women have an attachment to territory, as defined here, but owner-ship of land and the other social and power relationships which are included in the concept are unequal in almost all cases. This is illus-trated by a number of the case studies, perhaps most forcefully by Alice Nkhoma-Wamunza in Chapter 9. Cleavages and conflicts

241

within territorial units, which are based on factors such as gender, will have to be specifically considered. They mark a major difference between development from within and the older concept of development from below. They do not, however, negate the utility of territory as an integral part of development from within.

Territory can obviously be defined at a variety of scales but a precise definition of 'local scale' is a problem. Gore has argued that one scale must be considered as dominant because what is 'endogenous' at one scale can be 'exogenous' at another simply by changing the definition of region.

The definition of local scale will vary from place to place but it will usually be the smallest territory which is effective and efficient and which has meaning for the local people, who will define their own 'life space' in terms of their own values and realities.

Dele Olowu uses the term 'local self-governance' which has some similarities to development from within. He argues that local self-governance has 'three important attributes: locality, primary accountability to the local people, and the presence of important regulatory, economic or social services or a combination of all' (Olowu 1989: 205). He argues that 'smallness/localness' is one of the characteristics which give local self-governance its logic and dynamic and he presents evidence that the arguments of economies of scale in support of the centralization of local government service are overstated. He offers interesting quantitative evidence on the average population for existing local government units in Africa which shows that in only one case, Zimbabwe with an average of 6,000 people per unit, are units of a size which would encourage 'localness'. The figures for other countries in which the case studies in this book are situated are 136,090 (Kenya), 166,386 (Tanzania) and 187,692 (Ghana). In comparison, the figures for France (1,320), West Germany (2,634), USA (2,756), Italy (2,717) and Canada (6,372) are small.

After World War II, the British colonial government instituted a system of local government which marked a departure from the policy and practice of indirect rule. Olowu argues that these systems were remarkably effective because they were efficient, democratic and local. 'In addition the self-governance principle was closer to pre-colonial systems of community governance in most parts of Africa' (Olowu 1989: 206). These flourished for a relatively brief period in the 1950s before independence, after which

242

the pendulum swung to the other extreme in which African governments, confident about the powers and potentials of central government departments to promote and mobilize development and anxious to integrate their politics and to consolidate the positions of the ruling elites through the elimination of all opposition to their administrations, abolished existing local government structures in favour of structures which were usually adorned with management or development committees whose deliberate powers were severely curtailed.

(Olowu 1989: 208)

Although the brief period of local governance in the last years of colonialism is controversial, and there will be considerable disagreement over Olowu's interpretation, the approach appears to have had advantages. An important point for proponents of development from within is that it was based on a territorial unit at the local scale, often that of the town or village in the Nigerian situation. The case study from Zimbabwe (Zinyama, Chapter 2) may owe some of its success to the local government structures in place there.

None of the case studies in the book explicitly discusses the territorial nature of development from within but territorial organization, a sense of belonging to a distinct community, is implicit in all of them.

Territory refers to both physical and social space. The importance of physical space, the environment in which people live, should not be underestimated. The climate, soils, topography and weather are major factors in determining the pace of development and the success of any development initiative. Zinyama draws explicit attention to this factor which is often underestimated in the social science literature.

Ecology and environment are key factors in the development process and the African environment is incredibly complex and diverse and is far from fully understood. The variations in the microenvironment are considerable and local knowledge of these microenvironments and how best to utilize them resides primarily at the local level. There is a large literature on the importance of indigenous knowledge (Atte 1989; McCall 1988; Richards 1985); and an important element of development from within is the utilization of that knowledge. Such knowledge is often territorially specific.

Although the environmental problems of poverty are quite different from the environmental problems of surplus that are facing post-

243

industrial societies, it is erroneous to think of African societies as being unaware or ill-informed of the need to practise sustainable development. The case study materials show an awareness and concern for the environment and a deep understanding of it. This does not mean that African people have not put their environments under pressure because there is evidence that this does happen. However, this is usually out of necessity rather than choice. Chambers (1989a) edited a very interesting issue of the IDS Bulletin on 'Vulnerability: how the poor cope' which considers some of these issues. In that issue, de Waal discusses the famine situation in Darfur in Sudan and reports that:

> Famine victims in Darfur held that they had no control over their chances of dying during the famine. They did believe that they had power over their chances of preventing destitution, that is, preserving the basis of their livelihood. Consequently, their primary aim during the famine was not to minimise the probability of dying but to keep their animals alive and to cultivate.
>
> (de Waal 1989: 67)

Even under extreme pressure and in the face of death, this particular society was concerned with sustainability.

Territory, as defined for the purposes of development from within, also includes the social relationships of the community inhabiting the physical space. Rural communities are far from homogeneous entities; there are many different actors involved and the tensions and cleavages which exist need to be explicitly considered (McCall 1988). These include what Watts has called 'the rough and tumble of peasant political-economy: namely indebtedness, deleterious terms of trade, gender oppression, local class structure, differential access to and control over resources' (Watts 1986: 382). The case studies reveal ample evidence of this. Songsore's study of North-West Ghana (Chapter 4) is an excellent example of all of the issues listed by Watts, as is Manuh's study from Ada District in southern Ghana (Chapter 5). Alice Nkhoma-Wamunza (Chapter 9) provides evidence of gender oppression, while gender relationships are also a key element in Chavangi's study from Kakamega District (Chapter 7). Gender relationships clearly need particular attention and it has been argued that, in order to improve our understanding, we need to 'pry open the black box of the household'. Sarah Berry argues that:

244

Rather than taking the household as a unit of analysis we need to treat it as a point of departure . . . The questions we need to ask are not how do households decide but rather how does membership in a household affect people's access to resources, obligations to others and understanding of their options.

(Berry 1984a: 23)

Harriet Friedman points out that 'Family enterprise is a battleground over patriarchy where property is immediately at stake' (Friedman 1986: 192), which is especially significant in relation to access to and ownership of land, a critical resource for rural people.

African societies must be understood in their own context. This is by no means static and the argument is perhaps best placed in terms of the co-existence of community cohesion and the tension of different sub-community interest groups which varies over time and space.

The case study by Dei (Chapter 3) and a number of the additional case studies cited show that communities often act in a co-ordinated collective manner. Many African rural societies are organic entities which operate simultaneously at various levels with community cohesion and sub-community tensions existing side by side. Local initiatives at the sub-community level, such as the activities of women's groups, for example, are often organized and implemented in a manner which attempts to avoid confrontation. Contradictions are rarely resolved through confrontations or coercion but rather through negotiation (Watts 1988). Confrontation does take place, as in the cases of the Ada Songor Salt Miners' Co-operative Society (Manuh, Chapter 5) and the Utengule Usangu brewers (Nkhoma-Wamunza, Chapter 9), but this appears to be a last resort when other approaches have failed. Development from within must, however, explicitly recognize and consider the realities of rural society as opposed to the mythology and must also accept that the community is a dynamic and changing entity in many cases.

Development from within argues for the maximum utilization of the resources of a territory primarily for the satisfaction of the inhabitants of that territory. This includes both the physical and human resources of the local community. It is true that many communities are poor in both absolute and relative terms. In cases such as North-West Ghana (Songsore, Chapter 4), for example, both the resource base and the human base have been eroded by induced or forced migration, extraction and exploitation. But it is also true that in many local communities there are resources which remain underutilized or unutilized. The case studies in this book and those cited in the literature show that, in both human and material terms, much can be achieved by liberating or reviving the enthusiasm and knowledge of local people. In this sense, development from within is a self-reliant concept although, as will be discussed later, it is not an autarchic concept. Relationships with the State, and possibly involvement with third parties such as non-government organizations, are an important consideration for development from within. Selective spatial closure, as described in the original concept of development from below, is still an option for development from within but only in very exceptional circumstances.

PROSPECTS FOR DEVELOPMENT FROM WITHIN

Development from below was described by some authors as utopian. At the time when *Development from above or below?* (Stöhr and Taylor 1981) was written, the political and economic climate in Africa and the mainstream of development ideology was such that the development from below paradigm received little policy acceptance. A decade later, the situation has changed somewhat and the possibilities of a revised and revitalized development from within approach are greater. This does not mean that the paradigm will be easily accepted or that there are still no contradictions but the likelihood of acceptance has increased. It is ironic and unfortunate that the spiral of economic decline which has led to the current crisis in Africa has in itself created the preconditions in which development from within is more likely to be acceptable to local people, African governments, aid agencies, non-government organizations and major international funding agencies such as the World Bank. The crisis has also led to a reappraisal of both ideologies and strategies of development (Chambers 1989b).

From the perspective of the local people has come the realization that they cannot depend on the State either to meet their basic needs or to further their development objectives. As a result, they are taking their destiny into their own hands to a much larger extent than ever before. The realities of the last decade are eroding what is left of the 'colonial mentality' and the dependence on government as the main provider of services and economic opportunities. There are exceptions to this, such as the case of Zimbabwe (Zinyama, Chapter 2), but the emergence of local coping strategies in increasing numbers throughout the continent supports this view.

There is general acceptance that African governments, with few exceptions, have been unable to meet their development goals. As a result, either by choice or by necessity, the State is in a sense 'withdrawing' from many of the rural areas (Dearlove and White 1987), at least in terms of the provision of adequate development services, although control functions continue to exist. Several states are involved in Structural Adjustment Programmes, which are based on a resurgence of neo-classical macroeconomic theories, under the guidance of the World Bank. Structural adjustment often involves a severe cut-back in services to rural people. In these circumstances there is some 'space' for development from within to emerge and develop. The control functions of the State continue to exist and it can be argued that in some countries there has been a move away from democratization and a climate of increased repression associated with structural adjustment programmes. But unless the State perceives itself to be directly challenged at the political level it appears to have neither the resources nor the inclination to intervene as it would have a decade ago. Non-government organizations and to a lesser extent aid agencies have policies which favour working at the grass-roots level. Again, this strengthens the opportunities for development from within. Although NGOs cannot fill all of the 'development space' that is left by the State, they can often supply support to development from within initiatives in a much more promising way than the State. The thinking of the World Bank is in favour of 'market forces' performing functions which were previously carried out by the State. There is unlikely to be any objection from the Bank to the emergence of development from within although the reasons for this acceptance may well be quite different from those of the authors writing in this book.

Finally, there is increasing acceptance in academic circles of the need for a 'strategy of reversals' (Chambers 1989b) of which development from within is a good example. Crisis and the changes in attitudes, responses and policies that it appears to have encouraged has created the necessary preconditions for development from within.

DEVELOPMENT OR SURVIVAL

The UNCRD/CIRDAFRICA seminar in Arusha in 1989, which examined a number of case studies on local strategies and initiatives, showed that 'there were numerous individual and group (communal) initiatives, strategies and responses' (UNCRD 1989: 13). The empirical evidence that is presented in this volume and in a number of the other studies cited in the text provides additional proof of that.

A major conclusion of the UNCRD/CIRDAFRICA seminar was that there were two different types of strategies: one which was categorized as 'development oriented' and the second primarily as for 'survival'.

> In some countries . . . communal and social responses had been given major importance and support; hence successful local strategies had emerged through specific local organisations and institutions along with major government and NGO support. [In other countries, however,] the individual responses of rural people had prevailed to a greater extent, especially in cases of stress situations or conditions. The essential distinction between these two types of responses was that, in the former case, local strategies were long lasting and primarily development oriented, while in the latter case they reflected short-term and primarily survival needs.
>
> (UNCRD 1989: 13–14)

This is an interesting and important distinction which requires careful reflection. It is similar in some ways to the distinction between 'strategic' and 'survival' initiations that was mentioned earlier (Mackenzie, Chapter 1). 'Strategic' initiatives are those which lead to impact at scales larger than the local, and which begin to alter power relations at the national scale. This is also a point which is stressed by Goulet in his consideration of participation. He argues that

> the most difficult form of participation to elicit and sustain is also the most indispensable to genuine development. This is the type of participation which starts at the bottom and reaches progressively upward into ever widening areas of decision making . . . It matures into a social force building a critical mass of participating communities now enabled to enter into spheres of decision or action beyond their immediate problem solving.
>
> (Goulet 1989: 168)

Is a distinction between local 'development' initiatives and local 'survival' initiatives valid? Ndaro (Chapter 8), for example, considers many of the initiatives that are taken by the people of Dodoma District to be marginal, as they either are not integrated into government development plans or take different directions from what government would like. He describes as 'survival' those activities such as trade, crafts, industries, the construction of small dams and the bringing of new land under cultivation but, as was argued earlier, this distinction is dubious because it could equally well be interpreted as an indication that the people did not share the enthusiasm of the government for building roads, schools and hospitals, which Ndaro considers 'development'.

If development is measured by the criteria that were outlined at the beginning of this chapter then the distinction between 'development' and 'survival' becomes more blurred. If 'survival' strategies meet basic needs, as seems to be the case from much of the empirical evidence, is it valid to discuss them as marginal or not worthy of the term 'development'? The people of Ayirebi, for example (Dei, Chapter 3) are increasing economic growth, are concerned with equity and distribution of the fruits of that growth, have some control of their own destiny and appear to be achieving or returning to the traditional transcendental values which are held by the community to be of value. They do not appear to be influencing national policy or power relations in any significant way or 'entering into spheres of decision or action beyond their immediate problem solving' (Goulet 1989: 168). In fact, it might be argued that, by remaining 'invisible' to the State and other powerful actors in Ghana, they are improving their chances of controlling their own development.

Despite these arguments, there is value in considering distinctions between different types of local initiatives but more in order to determine the characteristics which have led to success than to develop typologies.

When the empirical evidence is examined, key elements for success appear to be the development of specific local organizations and institutions, which are controlled by the people themselves, and a degree of external support from an NGO, powerful 'patrons' or even government itself. Another element appears to be a lack of any direct and open challenge to the political power of the State. Bratton argues that what is needed is 'voice', which he sees as quite different from 'empowerment', a concept which he considers 'is used with rhetorical abandon but with little analytical precision in the literature' (Bratton 1990: 95). He argues further that

If 'empowerment' is taken to mean the capacity to get what you want, it is certainly an achievable goal in a local context where community-based organisations can sometimes effectively challenge a traditional power structure . . . But the concept of

'empowerment' exaggerates and raises unfulfillable expectations in the context of national politics in Africa. To try openly to
rearrange the allocation of resources on which the power
structure of the state rests is to run the risk of political repression.
(Bratton 1990: 95)

He uses the example of the National Farmers' Association of
Zimbabwe to illustrate the success of a preferred approach. This
organization meets most of the conditions under which Bratton
argues that 'organisations representing the rural poor are able to
attain a modicum of policy influence and to alter the allocation of
public resources' (Bratton 1990: 114). These include: the cultivation
of formal and informal ties with major political actors; a homogeneous and cohesive membership with membership accountability
at the local level; a federated structure to increase its policy voice;
specialization in policy issues of limited scope; and a reliance on
domestic rather than foreign funding. There may well be considerable disagreement with Bratton over these points but the empirical
evidence is clear; open confrontation with government, which is
perceived as a threat to the government's political power, leads to
repression. Sustainability is a key measure of success and if a local
initiative is destroyed then little has been gained.

There is, of course, a very fine line to be trod. There is an increasing danger of local initiatives and organizations being captured or co-
opted by government if they come too close. As support of local
initiatives and participation becomes increasingly more fashionable
in development circles, the autonomous organizations of the rural
poor, which were set up by the participants to meet their own needs in
ways defined by and acceptable to the community, may be over-
whelmed by outside influences and support and, in the process, may
be radically changed in character. In many situations local initiatives
succeed because they remain 'invisible' to those outside of the society.
The 'grey' or parallel economy often fares better in the shadows.
Rural people have learned from experience to keep some of their
activities hidden, especially from government, which has a tendency
to control, regulate or tax such initiatives. Bratton (1989) cites the
example of the Savings Development Movement in Zimbabwe,
which is described at the local level by Zinyama in Chapter 2.

Unfortunately, the very success of the savings movement in
Zimbabwe attracted so much attention that the government
found it necessary to intervene . . . By 1987 the government

was able to force changes in the governance of the savings movement by installing their own appointees on the Board of a reconstituted and renamed Self-Help Development Foundation.

(Bratton 1990: 98–9)

The study by Kobiah (1985) of local initiatives in Mathare Valley in Nairobi, although in an urban setting, is another example. The squatter co-operatives which were established by the residents in order to buy the land on which their shanty dwellings stood, in an effort to guard against government raids and evictions, were rapidly taken over by private companies speculating in land. The indigenous local initiatives gradually lost their dynamism as the three major external actors, the government of Kenya, the Nairobi City Council, and the National Christian Council of Kenya, became increasingly significant in decision-making.

Local initiatives are often circumscribed by lack of material resources and by poverty, which are the very conditions that they were formed to overcome. To be sustainable and to grow over time they need access to financial and other resources from outside the community. How to create conditions which allow rural peoples to improve their own lot without losing control of their own initiatives and institutions is an important part of development from within.

RELATIONSHIPS BETWEEN THE STATE AND LOCAL COMMUNITIES

The relationships between the local community and the State have been touched on earlier in this chapter as well as in almost every chapter in the book. These relationships are an extremely popular topic in the current development administration and political science literature on Africa. It is also key to the prospects for development from within.

As the State continues to be unable or unwilling to deal with the current challenges of African development, or implements policies which are seen by rural peoples to be detrimental to their interests, the significance of local organization and initiative increases. However, as both Brett and Beckman point out, the role of the State is still critical and it would be a serious error to underestimate this. Brett believes that there is considerable potential to reform current State practices and argues that 'the role of the state must remain central in any effective development strategy' (1988: 4). He discusses a number of solutions for increasing flexibility on the part of the State and his observations on what he calls 'co-operative and voluntary alternatives' (Brett 1988: 10) are worthy of careful consideration in relation to the context and rationale for local initiatives. He points out that 'collective organisation requires greater degrees of overall competence than autocratic management' and that 'Voluntary organisation can also transfer the cost of providing services from the wealthy taxpayer to the poorest members of the community and increase inequalities in access to services' (Brett 1988: 10). He goes on to observe that

> these methods [i.e. voluntary organization] can improve service provision and allow local groups to maintain their autonomy against powerful private interests and a distant and perhaps alien state. In most cases success will probably only occur where they are also given substantial degrees of support by state and aid agencies, particularly in the early stages before they have built up their resources and skills. Thus it is almost certainly the case here, as elsewhere, that the full development of autonomous agencies is not a substitute for effective public provision, but only possible where it can grow effectively in association with it.
>
> (Brett 1988: 10)

Beckman cautions that the domestic ruling classes must be taken seriously and not simply dismissed as corrupt, bureaucratically rigid, neopatrimonial compradors. He points out that

> There is also a commitment to the advancement of the nation, as perceived, no doubt, from a particular class perspective. There is a profound resentment of a heritage of foreign domination and racial arrogance, a desire to remove the stigma of inferiority and to seek a rightful place in the community of nations. [This is] a hungry and ambitious class [who] represent a dynamic force geared to expansion rather than decline.
>
> (Beckman 1988: 28)

The current crisis, far from leading to a decline or demise of State power, simply 'hastens historical processes . . . whereby local ruling classes, pampered by state protection, are pushed into deeper water and made to swim' (Beckman 1988: 28). An example might be the steps that were taken by the Nigerian Government, as described by

Mabogunje. These comprise a *domestically* initiated Structural Adjustment Programme which involves 'an agonising reappraisal of the true course of any development effort' (Mabogunje 1988: 17) and the creation of a powerful, new Directorate of Food, Roads and Rural Infrastructure to promote grass-roots social mobilization. If local initiatives are to succeed then it is clear that new power-sharing relationships must be worked out with powerful State actors. External agents, such as international agencies and NGOs, may have a mediating role to play in this process.

Robert Chambers presents the concept of the 'enabling state', which is part of his overall concept of a strategy of reversals, as described earlier. He argues that the paradigm of reversals 'resolves the contradiction between the neo-Fabian thesis that the state should do more, and the neo-liberal antithesis that the state should do less' (Chambers 1989b: 20). The task for Chambers is to dismantle the 'disabling state' and to replace it with 'a state which is not only protector and supporter, but also enabler and liberator' (Chambers 1989b: 20).

Chambers argues that there are three functions of the State which are fundamental for the rural poor: maintaining peace and the democratic rule of law; providing basic infrastructure and services; and managing the economy. He also outlines an agenda for abstention and an agenda for action. Decentralized process and choice are central to his paradigm of reversals and 'In this mode the state is not school but cafeteria, and development is decentralised, becoming not simpler but more complex, and not uniform but more diverse' (Chambers 1989b: 20).

His analysis and suggestions in respect to relations between the State and local community fit well with the concept of development from within. Despite the difficulties, the political climate exists to make development from within possible. There is 'political space' available to allow local communities to establish new relationships between themselves and central government. As Bratton observes, 'the reach of the African state routinely exceeds its grasp' (Bratton 1989: 585) and in the current circumstances the ability of the State to regulate and control has been reduced.

NON-GOVERNMENT ORGANIZATIONS

From the empirical evidence it is clear that the role and significance of non-government organizations has increased. One author has gone so far as to suggest that 'the 1980s may be known as the development decade of non-government organizations (NGOs)' (Fowler 1988: 1), a view which is echoed by Bratton (1989) who states that

> NGOs have entered the limelight as governments throughout Africa have begun to retreat from ambitious attempts to sponsor socioeconomic development 'from above' . . . Especially in the remotest regions of the African countryside, governments often have had little choice but to cede responsibility for the provision of basic services to a church, an indigenous self-help group, or an international relief agency.
>
> (Bratton 1989: 569)

NGOs have a potential comparative advantage over government in micro-level development but 'you cannot rely on all NGOs in all circumstances to be more effective; being an NGO is not synonymous with being better than government as an agent of micro-development' (Fowler 1988: 23). In some instances, if that potential comparative advantage is not realized, 'The current infatuation of official donors with NGOs will turn into disillusionment, and in the 1990s observers will be writing the obituaries of NGOs in micro-development' (Fowler 1988: 22).

The role of NGOs as perceived by government may be quite different from the perception of their role by local communities. There is congruence in terms of the needs of both people and government for increased resources and the provision of much needed services but the local people appear to view the NGOs as protectors and advocates whereas the government appears to view them as a 'development platform' for activities which they would like to see carried out but either will not or cannot finance from their own resources.

The empirical evidence shows that in many instances where local initiatives have been sustained and expanded there has been some involvement from the outside. The scale and nature of this involvement has varied but it appears to have been important, especially in facilitating the organizational side of local community initiatives and in providing the initial resources to allow these initiatives to be 'jump started'. In some instances the NGOs have been local, in others national, and in yet others international. In both Nigeria and Ghana, men, and to a lesser extent women, who have left their home village or town have been important players in furthering the welfare of their home communities both by remitting resources and by action on behalf of the community in the wider political and bureaucratic arena (Olowu 1989).

As with government, there is a danger that an NGO, especially one which is based in another country or one which is, say, religion-based and which has a hidden agenda of its own, may subvert or dominate the initiatives of local communities. There is indeed evidence in the case studies cited of this happening. In general, however, the empirical evidence which is available shows that NGOs have been increasingly supportive of local initiatives.

Both Fowler (1988) and Bratton (1989, 1990) explore these matters more fully and the NGOs themselves have issued an interesting Declaration on the African Economic and Social Crisis (UN 1986b). In the emerging new relationship between the local community and the State, the NGOs may have a particularly significant role to play and may act as a catalyst of development from within.

POLICY ACTIONS

Development from within is not a normative concept and, since the major actor in the whole process is the local community itself, it is difficult, if not impossible, to suggest policy actions at the most critical decision-making level. However, it is possible to look at policies which can be adopted by the other major actors, such as government, non-governmental agencies and international funding agencies. The principal conclusions of the UNCRD/CIRDAFRICA seminar provide an excellent synopsis of what is required:

> Given the situation that has been prevalent in the region and the local realities, local initiatives offer great scope for promoting self-sustaining local development and, through it, national development.
>
> It was, therefore, seen as imperative that governments, international agencies, and the NGOs concerned provide a conducive environment and support for the promotion of such initiatives.
>
> (UNCRD 1989: 28)

A number of specific suggestions were made. The role of government was considered 'as crucial in providing a suitable social, economic and political environment in which local initiatives and local self-development could flourish' (UNCRD 1989: 21). The State, therefore, must maintain peace, stability, law and order; provide basic socioeconomic infrastructure and management of the economy; provide a policy framework to encourage local initiatives, including official recognition of local initiatives and institutions as an integral part of the national development effort; provide financial, institutional and technical support for the activities of local institutions and initiatives; decentralize decision-making and adopt a regional or local development strategy in which local resources and institutions would be the major platform of development for the region; and establish a national institution to co-ordinate and link local organizations, initiatives and strategies. The name that was suggested was the National Indigenous Resource Centre or the National Centre for Local Self-Reliance.

> This institution should act as an organiser, facilitator and catalyst, and provide avenues for assistance to local initiatives in terms of funds, technical support, supplies, training and management advice. It could work with communities to modernise local technological capabilities or indigenise appropriate outside technologies. On top of this, it should compile, document and publicise successful local initiatives and strategies (both technical and otherwise) nationwide, to act as incentives to all local communities. Local people across the nation should be appointed members of this national institution.
>
> (UNCRD 1989: 23)

In some ways these suggestions echo those of Chambers (1989b), which were described earlier in this chapter. He argues that the State has done too much in some areas and much too little in others. The role of the NGOs is also considered important in encouraging local initiatives. The UNCRD/CIRDAFRICA seminar suggested: closer co-ordination and less competition among NGOs; that NGOs provide technical, financial, and other information for local initiatives together with management and organizational training; and that the work of NGOs in local communities be guided by a number of basic principles which ensure that the local people are in full control of decision-making and that existing local structures and organizations are respected and built on rather than replaced.

Many of the policy suggestions for NGOs apply to international funding agencies too but it was also suggested that they concentrate on institutional support; finance and design the improvement of local technology rather than the purchase of machinery from abroad; direct efforts to programmes such as public health care and disease eradication which improve the local environment and make it more conducive to the emergence of local initiatives; and to help to fund research and implementation of local technologies, agricultural systems, manufacturing, etc. which are of direct concern to local people.

Development from within is not primarily a planning strategy but the policy implications of such an approach cannot be ignored if change is to take place.

There are clearly obstacles to be overcome and certain preconditions in which development from within is more likely to emerge. These include: 'the existence of a political will that recognise[s] the role of local initiatives and organisation in local and national development and promote[s] such organisation' (UNCRD 1989: 26); the provision of incentives to local people to meet their own development needs; decentralization and devolution of the structure of government; education and sensitization of local communities to understand the nature of the realities facing them; financial resources and institutions which are geared to the needs of local communities and organizations; training opportunities to increase managerial and administrative skills of local people; and recognition of the potential of local communities, including women, in the solution of their own development problems. In a sense this turns existing development approaches completely on their head.

CONCLUSION

The human potential, basic wisdom and knowledge of Africa's local peoples have been seriously underestimated. To some, they are seen as a major barrier to the development aims both of the African state and of the international agencies which are involved in promoting and guiding what they see as being required for development to take place. They are objects of development who have to be 'modernized', 'mobilized' or 'captured'. A basic objective of development from within is to allow local people to become the subject, not the object, of development strategies. Given the opportunity to do so, they have shown themselves to be perfectly capable of making rational choices regarding their own destinies. Too much attention has been given to the negative developmental aspects of issues such as 'kinship' and not nearly enough to the positive aspects of local community realities on which a more meaningful development can be built. It is hoped that this book has helped to contribute to an increased understanding of a development from within for, by and with African peoples, as a result of which a new optimism for Africa's future may emerge from the current pessimism.

BIBLIOGRAPHY

Adedeji, A. (1985) 'The Monrovia Strategy and the Lagos Plan of Action: five years after', pp. 9–34 in A. Adedeji and T.M. Shaw (eds) *Economic crisis in Africa: African perspectives on development problems and potentials*, Lynne Rienner, Boulder, Col.

Adedeji, A. (1987) 'A preliminary assessment of the performance of the African economy in 1986 and prospects for 1987', End-of-Year Conference at ECA Headquarters, 2 January 1987, Addis Adaba.

Adedeji, A. (1989) 'The African Alternative Framework to Structural Adjustment', unpublished paper, presented at the University of Ottawa, Canada, 23 October 1989.

Agricultural Marketing Authority (1988) *Cotton situation and outlook report (annual)*, Harare.

Akator, G. (1988) *Youth Development Task Force, weekend seminar report*, Dodoma Municipal Council, Dodoma.

Alila, P.O. (1988) 'Rural development in Kenya: a review of past experience', *Regional Development Dialogue* 9(2): 142–65.

Anim, N.O. (1959) 'A local study of a coastal district', *Ghana Geographical Association Bulletin* 4(2): 16–21.

Annegers, J.F. (1973) 'Seasonal food shortages in West Africa', *Ecology of Food and Nutrition* 2: 251–7.

Arhin, K. (1983) 'Peasants in 19th century Asante', *Current Anthropology* 24(4): 471–9.

Asante, S.K.B. (1985) 'Development and regional integration since 1980', pp. 79–99 in A. Adedeji and T.M. Shaw (eds) *Economic crisis in Africa: African perspectives on development problems and potentials*, Lynne Rienner, Boulder, Col.

Asgele, Tsegay (1989) 'Case study on local people's rural development strategies and responses in the Welo Region of Ethiopia', Paper presented to the UNCRD/CIRDAFRICA Seminar on Reviving Local Self-Reliance: Challenges for Rural/Regional Development in Eastern and Southern Africa, Arusha, Tanzania, 24 February.

Atsu, S.Y. (1984) 'The effect of government policies on the increased production of food and food self-sufficiency in Ghana', Paper read at the International Conference on Food Self-sufficiency in West Africa, University of Ghana, Legon, Accra, 1–3 May.

259

Atte, David (1989) 'Indigenous local knowledge as a key to local-level development: possibilities, constraints and planning issues in the context of Africa', Paper presented to the UNCRD/CIRDAFRICA Seminar on Reviving Local Self-Reliance: Challenges for Rural/Regional Development in Eastern and Southern Africa, Arusha, Tanzania, 24 February.

Bassand, Michel, Brugger, Ernst A., Bryden, John M., Friedmann, John and Stuckey, B. (eds) (1986) *Self-reliant development in Europe*, Gower, Aldershot.

Beckman, B. (1988) 'The post-colonial state: crisis and reconstruction', *IDS Bulletin* 19(4): 26–34.

Bell, M. (1979) 'The exploitation of indigenous knowledge, or the indigenous exploitation of knowledge: whose use of what for what?', *IDS Bulletin* 10(2): 44–50.

Belshaw, D. (1979) 'Taking indigenous technology seriously: the case of inter-cropping techniques in East Africa', *IDS Bulletin* 10(2): 24–8.

Bentil, B., Gadway, J., Huttenrauch, H., Monikes, V. and Schmidt, R.H. (1988) *Rural finance in Ghana (a research study on behalf of Bank of Ghana)*, September, Accra.

Bequele, A. (1983) 'Stagnation and inequality in Ghana', pp. 219–47 in D. Ghai and S. Radwan (eds) *Agrarian policies and rural poverty in Africa*, ILO, Geneva.

Bernstein, H. (1979) 'African peasantries: a theoretical framework', *Journal of Peasant Studies* 6(4): 421–43.

Berry, S. (1984a) *Households, decision making and rural development*, Development Discussion Paper no. 167, Harvard Institute of International Development, Cambridge, Mass.

Berry, S. (1984b) 'The food crisis and agrarian change in Africa: a review essay', *African Studies Review* 27(2): 59–112.

Berry, S. (1989) 'Social institutions and access to resources', *Africa* 59(1): 41–55.

Bienefeld, M. (1989) 'Structural adjustment and rural employment in Tanzania', Unpublished paper submitted to International Labour Office Employment/Rural project on structural adjustment and rural labour markets in five African countries.

Blaikie, P. (1989) 'Environment and access to resources in Africa', *Africa* 59(1): 18–40.

Boateng, E.A. (1959) *A geography of Ghana*, Cambridge University Press, Cambridge.

Boateng, I.K. (1986) 'Governmental and voluntary participation in Ghana's rural development programme', pp. 27–47 in Brown, C.K. (ed.) *Rural development in Ghana*, Ghana Universities Press, Accra.

Bradley, P.N. (1984) *The District Resource Analysis as applied to Kakamega*, Working Paper 1, Kenya Woodfuel Development Programme, Nairobi.

Brandtzaeg, B. (1982) 'The role and status of women in post-harvest food conservation', *Food and Nutrition Bulletin* 4(11): 33–42.

Bratton, M. (1986) 'Farmer organizations and food production in Zimbabwe', *World Development* 14(3): 367–84.

Bratton, M. (1987) 'Drought, food and the social organisation of small farmers in Zimbabwe', pp. 213–44 in M.H. Glantz (ed.) *Drought and hunger in Africa: denying famine a future,* Cambridge University Press, Cambridge.

Bratton, M. (1989) 'The politics of government–NGO relations in Africa', *World Development* 17(4): 569–87.

Bratton, M. (1990) 'Non-governmental organizations in Africa: can they influence public policy?' *Development and Change* 21: 87–118.

Brett, E.A. (1987) 'States, markets and private power in the development world: problems and possibilities', *IDS Bulletin* 18(3): 31–7.

Brett, E.A. (1988) 'Adjustment and the state: the problem of administrative reform', *IDS Bulletin* 19(4): 4–11.

Brooke, C. (1967) 'The heritage of famine in Central Tanzania', *Tanzania Notes and Records* 67: 15–22.

Brown, C.K. (ed.) (1986) *Rural development in Ghana,* Ghana Universities Press, Accra.

Brugger, E.A. and Stuckey, B. (1986) 'Introduction: self-reliant development in Europe', pp. 1–8 in Bassand *et al.* (eds) *Self-reliant development in Europe,* Gower, Aldershot.

Bryceson, D.F. and Kirimbai, M. (1980) *Subsistence and beyond: money earning activities of women in rural Tanzania,* BRALUP Research Report 45, University of Dar es Salaam.

Callear, D. (1984) 'Land and food in the Wedza communal area', *Zimbabwe Agricultural Journal* 81(4): 163–8.

Campbell, David J. (1990) 'Community-based strategies for coping with food scarcity: a role in African famine early-warning systems', *Geojournal,* March, 231–41.

Carney, J. (1988) 'Struggles over crop rights and labour within contract farming households on a Gambian irrigation rice project', *Journal of Peasant Studies* 15(3): 334–9.

Central Bureau of Statistics (1982) *Census office report,* Institute of Scientific, Social and Economic Research, Legon, Accra.

Chale, F.U. and Ngonyani, G.H. (1979) *Report on a survey of co-operative income-generating projects for women and its impact on the social welfare of children and the family,* BRALUP Workshop on Women and Development Paper 24, Dar es Salaam.

Chambers, R. (1988) 'Bureaucratic reversals and local diversity', *IDS Bulletin* 19(4): 50–6.

Chambers, R. (1989a) 'Editorial introduction: vulnerability, coping and policy', *IDS Bulletin* 20(2): 1–7.

Chambers, R. (1989b) *The state and rural development: ideologies and an agenda for the 1990s,* IDS Discussion Paper 269, University of Sussex.

Chanock, M. (1985) *Law, custom and social order: the colonial experience in Malawi and Zambia,* Cambridge University Press, Cambridge.

Chavangi, N.A. (1988) Case Study of Women's Participation in Forestry Activities in Kenya, Kenya Woodfuels Development Programme, Nairobi.

Chavangi, N.A. and Ngugi, A.W. (1987) 'Innovatory participation in programme design: tree planting for increased fuelwood supply for rural households in Kenya', Paper presented at the workshop on Farmers and Agricultural Research, University of Sussex, July 1987.

Chavangi, N.A., Engelhard, R.J. and Jones, V. (1986) *Culture as the basis for implementing self-sustaining woodfuel development programmes*, Kenya Wood-fuel Development Programme, Nairobi.

Chileshe, Jonathan H. (1989) 'Socioeconomic structural change, forms of traditional authority and prospects for local-level development', Paper presented to the UNCRD/CIRDAFRICA Seminar on Reviving Local Self-Reliance: Challenges for Rural/Regional Development in Eastern and Southern Africa, Arusha, Tanzania, 24 February.

Chitsike, L.T. (1988) *Agricultural co-operative development in Zimbabwe*, Zimbabwe Foundation for Education with Production, Harare, August.

Christopher, A.J. (1971) 'Land tenure in Rhodesia', *South African Geographical Journal* 53: 39–52.

Colclough, C. (1988) 'Zambian adjustment strategy – with and without the IMF', *IDS Bulletin* 19(1): 51–60.

Collier, P. (1980) *Poverty and growth in Kenya*, World Bank Staff Working Paper 389, Washington, D.C.

Commonwealth Expert Group on Women and Structural Adjustment (1989) *Engendering adjustment for the 1990s*, Commonwealth Secretariat, London.

Cornia, G.A. (1988) 'Economic decline and human welfare in the first half of the 1980s', pp. 11–47 in G.A. Cornia, R. Jolly and F. Stewart (eds) *Adjustment with a human face: protecting the vulnerable and promoting growth*, Clarendon Press, Oxford.

Cotton Marketing Board (1988) *Annual report*, Harare.

Coulson, A. (1982) *Tanzania: a political economy*, Oxford University Press, Oxford.

CCA and CUA (1988) 'Credit for women', Ghana. Unpublished document.

Credit Unions Association (1971) *CUA Ed-1*, CUA Education/Information Department, Accra, April.

Credit Unions Association (1978) *A history of co-operative credit unions in Ghana, Tenth Milestone 1968–1978*, Accra.

Credit Unions Association (1985) CUA headquarters statistics for 1984/85 fiscal year.

Credit Unions Association (1987a) *Ghana co-operative Credit Unions Association (CUA) limited country progress report for the ACCOSCA biennial meeting and education conference*, Nairobi, 25–29 May.

Credit Unions Association (1987b) *Memorandum on reactivation of Ghana co-operative movement: Ghana co-operative Credit Unions Association (CUA) Limited*, Accra, 15 December.

Credit Unions Association (1988) *Model by-laws and articles of association for Ghana co-operative credit unions*, Accra.

Daddieh, C.K. (1985) 'Recovering Africa's self-sufficiency in food and agriculture', pp. 187–200 in A. Adedeji and T.M. Shaw (eds) *Economic crisis in Africa: African perspectives on development problems and potentials*, Lynne Rienner, Boulder, Col.

Daily Graphic (1989) 27 April, Accra.

Daily Graphic (1989) 21 July, Accra.

Daily Graphic (1989) 30 October, Accra.

Daily News (1989) 3 April, Dar-es-Salaam.

de Stemper, G.A. (1988) 'Progresso social: a mission to help women', *Credit Union World Reporter* 3(2): 8–10.

Dearlove, J. and White, G. (1987) 'The retreat of the State', *IDS Bulletin* 18(3): 1–3.

Dei, G.J.S. (1986) 'Adaptation and environmental stress in a Ghanaian forest community', Ph.D. dissertation, Department of Anthropology, University of Toronto, University Microfilms, Ann Arbor, Mich.

Dei, G.J.S. (1987) 'Land and food production in a Ghanaian forest community', *Africa Development* 12(1): 101–24.

Dickson, K.B. (1969) *A historical geography of Ghana*, Cambridge University Press, Cambridge.

Dutkiewicz, P. and Shenton, R. (1986) 'Debates', *Review of African Political Economy* 37: 108–16.

Dutkiewicz, P. and Williams, G. (1987) 'All the king's horses and all the king's men couldn't put Humpty-Dumpty together again', *IDS Bulletin* 18(3): 39–44.

Economist Intelligence Unit (1989) *Country report: Kenya* 1, London.

Evans, A. (1989) 'The implications of economic reforms for women in Zambia: the case of the Economic Reform Programme 1983–7', Unpublished paper commissioned for the Commonwealth Expert Group on Women and Structural Adjustment.

Evans, A. and Young, K. (1988) 'Gender issues in household labour allocation: the case of Northern Province, Zambia', ODA/ESCOR Research Report; cited in Evans (1989).

Floyd, B.N. (1962) 'Land apportionment in Southern Rhodesia', *Geographical Review* 52(4): 566–82.

Forbes, D.K. (1984) *The geography of underdevelopment*, Croom Helm, London.

Fortes, M. and Fortes, S.L. (1936) 'Food in the domestic economy of the Tallensi', *Africa* 9: 237–76.

Fowler, A. (1988) *Non-governmental organizations in Africa: achieving comparative advantage in relief and micro-development*, IDS Discussion Paper 249, Sussex.

Franke, Richard W. (1987) 'Power, class, and traditional knowledge in Sahel food production', pp. 257–81 in I.L. Markovitz (ed.) *Studies in power and class in Africa*, Oxford University Press, Oxford.

Friedman, H. (1986) 'Patriarchy and poverty', *Sociologie Ruralis* 26(2): 186–95.

Friedmann, J. (1981) 'Regional planning for rural mobilization in Africa', *Rural Africana* 12–13: 3–29.

Friedmann, J. (1986) 'Regional development in industrialised countries: endogenous or self-reliant', pp. 203–16 in Bassand *et al.* (eds) *Self-reliant development in Europe*, Gower, Aldershot.

Friedmann, J. and Weaver, C. (1979) *Territory and function: the evolution of regional planning*, Edward Arnold, London.

Galtung, J., O'Brien, B. and Prieswerk, R. (eds) (1980) *Self-reliance, a strategy for development*, Bogle L'Ouverture, London.

Gasper, D. (1988) 'Rural growth points and rural industries in Zimbabwe: ideologies and policies', *Development and Change* 19(3): 425–66.

Ghana (1984) *Population Census*, Government Press, Accra.

Ghana (1986) 'Amissah Committee of Enquiry Report', Accra, Unpublished document.

Ghana (1987) *Programme of Action to Mitigate the Social Costs of Adjustment*, Government Printer, Accra.

Glazier, J. (1985) *Land and the uses of tradition among the Mbeere of Kenya*, University Press of America, Lanham, Maryland.

Globe and Mail (1990) 6 July, Toronto.

Gobbins, K.E. and Prankard, H.A. (1983) 'Communal agriculture: a study from Mashonaland West', *Zimbabwe Agricultural Journal* 80(4): 151–8.

Gondwe, Z.S. (n.d.) 'Female interstate succession to land in rural Tanzania – whither equality?', Dar es Salaam Women Research and Documentation Project, Dar es Salaam.

Gore, C. (1984) *Regions in question: space, development theory and regional policy*, Methuen, New York.

Goulet, D. (1989) 'Participation in development: new avenues', *World Development* 17(2): 165–78.

Graham, Y. (1989) 'From GTP to Assene: aspects of industrial struggles 1982–7', pp. 43–72 in E. Hansen and K. Ninsen (eds) *The State, development and politics in Ghana*, Codesria Book Series, Codesria, Dakar, Senegal.

Grain Marketing Board (1988) *Annual report*, Harare.

Green, R.H. (1988) 'Ghana: progress, problematics and limitations of the success story', *IDS Bulletin* 19(1): 7–16.

Green, R.H. (1989) *Degradation of rural development: development of rural degradation – change and peasants in Sub-Sahara Africa*, IDS Discussion Paper 265, September.

Guyer, J.I. and Peters, P.E. (1987) 'Introduction: conceptualizing the household: issues of theory and policy in Africa', *Development and Change* 18(2): 197–214.

Havnevik, K.J. *et al.* (1988) *Tanzania country study and Norwegian aid review*, Centre for Development Studies, University of Bergen.

Herald, The (1989) 4 August, Harare.

Hewer, L. (1974) *Rural development: world frontiers*, Iowa State University Press, Des Moines.

HIARI (1982) Newsletter of the Community Development Trust Fund of Tanzania, 10–11.

Hill, P. (1963) *The migrant cocoa farmers of Southern Ghana*, Cambridge University Press, Cambridge.

Hill, P. (1975) 'The West African farming household', pp. 119–36 in J. Goody (ed.) *Changing social structure in Ghana*, International African Institute, London.

Hinderink, J. and Sterkenburg, J.J. (1987) *Agricultural commercialisation and government policy in Africa*, Routledge & Kegan Paul, London.

Hubbert, Lorrie (1988) 'Women's access to credit', *Credit Union World Reporter* 3(2): 15–17.

Hunter, J.M. (1967) 'Seasonal hunger in a part of the West African savannah: a survey of bodyweights in Nangodi, Northeast Ghana', *Transactions of the Institute of British Geographers* 41: 167–85.

Hutchful, E. (1990) 'The emperor in new clothes: new themes in the World Bank's concept of structural adjustment', Paper presented at the Annual Conference of the Canadian Association of African Studies Conference, Dalhousie University, Halifax, May.

Hyden, G. (1980) *Beyond Ujamaa in Tanzania: underdevelopment and an uncaptured peasantry*, Heinemann, London.

Hyden, G. (1983) *No shortcuts to progress: African development management in perspective*, University of California Press, Berkeley.

Igbozurike, M.U. (1971) 'Ecological balance in tropical agriculture', *Geographical Review* 61(4): 519–29.

Imoo, B.V. and Louse, J. (1988) 'Adult education programme Turkana District, Kenya', pp. 79–89 in P. Oakley (ed.) *Proceedings of the international symposium on the challenge of rural poverty: how to meet it*, German Foundation for International Development, Feldafing, FGR.

Jamal, V. (1988) 'Getting the crisis right: missing perspectives on Africa', *International Labour Review* 127(6): 655–78.

James, R.W. and Fimbo, G.M. (1973) *Customary land law of Tanzania: a source book*, East Africa Literature Bureau, Dar es Salaam.

Kalapula, E.S. (1989) 'Responses to drought in Namwala District: local community strategies and initiatives', Paper presented to the UNCRD/ CIRDAFRICA Seminar on Reviving Local Self-Reliance: Challenges for Rural/Regional Development in Eastern and Southern Africa, Arusha, Tanzania, 24 February.

Kapinga, N.B. (1989) 'Support to local self-reliant efforts: the approach of Rukwa Integrated Rural Development Project', Paper presented to the UNCRD/CIRDAFRICA Seminar on Reviving Local Self-Reliance: Challenges for Rural/Regional Development in Eastern and Southern Africa, Arusha, Tanzania, 24 February.

Karikari, S.K. (1971) 'Cocoyam cultivation in Ghana', *Legon Extension Bulletin* 13: 1–12, University of Ghana Legon, Accra.

Kauzeni, A.S. (1988) 'Rural development alternatives and the role of local-level development strategy: Tanzania case study', *Regional Development Dialogue* 9(2): 105–38.

Kennedy, E. and Cogill, B. (1987) *Income and nutritional effects of the commercialization of agriculture in southwestern Kenya*, International Food Policy Research Institute Research Report, No. 63, Washington, D.C.

Kenya (1982) *District focus for rural development*, Government Printer, Nairobi.

Kenya (1986a) Sessional Paper No. 1, *Economic management for renewed growth*, Government Printer, Nairobi.

Kenya (1986b) *Development Plan 1984–9*, Government Printer, Nairobi.

Kenya (1987) *Statistical abstract*, Government Printer, Nairobi.

Kenya (1989a) *Economic survey*, Government Printer, Nairobi.

Kenya (1989b) *Development plan 1989–1993*, Government Printer, Nairobi.

Kenya, Machakos District (1987) *Annual report*, Ministry of Culture and Social Services, Nairobi.

Kenya, Machakos District (1988) *Ministry of Co-operatives report*, Nairobi.

Kenya, Machakos District (1989a) *Machakos District development plan 1989– 1993*, Ministry of Finance and Planning, Nairobi.

Kenya, Machakos District (1989b) *Machakos District development plan annex*, Nairobi.

Kidd, R. (1982) 'Popular theatre and popular struggle in Kenya: the story of the Kamiriithu Community Educational and Cultural Centre', *Theaterwork* 2(6): 47–61.

Kinsey, B.H. (1983) 'Emerging policy issues in Zimbabwe's land resettlement programmes', *Development Policy Review* 1(2): 163–96.

Kitching, G. (1980) *Class and economic change in Kenya: the making of an African petite bourgeoisie*, Yale University Press, New Haven, Connecticut.

Kitching, G. (1985) 'Politics, method and evidence in the "Kenya Debate" ', pp. 115–51 in H. Bernstein and B.K. Campbell (eds) *Contradictions of accumulation in Africa*, Sage, Beverly Hills.

Kjekshus, H. (1977) *Ecology control and economic development in East African history: the case of Tanganyika, 1850–1950*, University of California Press, Berkeley.

Kobiah, S.M. (1985) 'The origins of squatting and community organization in Nairobi', *African Urban Studies*, special issue 19–20.

Koda, B., Mbliniya, M., Muro, A., Kokubdwa Nkebukwa, A., Nkhoma, A., Tumbo-Masabo, Z., and Vuorela, U. (1987) *Women's initiatives in the United Republic of Tanzania: a technical co-operation report*, World Employment Program, ILO, Geneva.

Kropp-Dakubu (1988) 'Multilingualism in Ada District', *Research Review* 4(1): Institute of African Studies, Legon, Accra.

Lardner, G.E.A. (1985) 'Beyond the neocolonial nexus: inheritance, implementation, and implications of the Lagos Plan of Action', pp. 35–46 in A. Adedeji and T.M. Shaw (eds) *Economic crisis in Africa: African perspectives in development problems and potentials*, Lynne Rienner, Boulder, Col.

Lenin, V.I. (1978) *Imperialism, the highest stage of capitalism*, Progress, Moscow.

Leys, C. (1987) 'The state and the crisis of simple commodity production in Africa', *IDS Bulletin* 18(3): 45–9.

Longhurst, R. (1988) 'Cash crops, household food security and nutrition', *IDS Bulletin* 19(2): 28–36.

Longhurst, R. Kamara, S. and Mensurah, J. (1988) 'Structural adjustment and vulnerable groups in Sierra Leone', *IDS Bulletin* 19(1): 25–30.

Loutfi, M. (1989) 'Development issues and state policies in sub-Saharan Africa', *International Labour Review* 128(2): 137–54.

Loxley, J. (1988) *Ghana: economic crisis and the long road to recovery*, North–South Institute, Ottawa.

Loxley, J. (n.d.) 'Structural adjustment programs in Africa: some issues of theory and policy' (draft).

Mabogunje, A. (1988) 'Africa after the false start', Paper presented to the 26th Congress of the International Geographical Union, Sydney, Australia.

McCall, M. (1987) 'The implications of East-African rural social structure for local-level development: the case for participatory development based on indigenous knowledge systems', Paper presented at seminar in Eastern Africa rural development experience: strategies in local level development, 30 June to 3 July 1987, Nairobi, UNCRD/IDS, pp. 1–36, Nagoya, Japan.

McCall, M. (1988) 'The implications of Eastern African rural social structure for local-level development: the case for participatory development based on indigenous knowledge systems', *Regional Development Dialogue* 9(2): 41–69.

Macebo, L. (1988) 'Vusanami collective farming cooperative society (VCCSL) Zimbabwe', p. 80 in P. Oakley (ed.) *Proceedings of the international symposium on the challenge of rural poverty: how to meet it*, German Foundation for International Development, Feldafing, FGR.

Mackenzie, F. (1986) 'Local initiatives and national policy: gender and agricultural change in Murang'a District, Kenya', *Canadian Journal of African Studies* 20(3): 377–401.

Mackenzie, F. (1990) 'Gender and land rights in Murang'a District, Kenya', *Journal of Peasant Studies* 17(4): 609–43.

Mackenzie, F. (1991) 'Political economy of the environment, gender and resistance under colonialism: Murang'a District, Kenya, 1910–1950', *Canadian Journal of African Studies* 25(2).

Mackenzie, F. and D.R.F. Taylor (1989) 'Inequality in Murang'a District, Kenya: local organization for change', pp. 112–39 in K. Swindell, J.M. Baba and M.J. Mortimore (eds) *Inequality and development: case studies from the third world*, Macmillan, London.

Manoukian, M. (1964) *Akan and Ga-Adangme Peoples*, Oxford University Press, Oxford.

Maro, P.S. (1990) 'The impact of decentralization on spatial equity and rural development in Tanzania', *World Development* 18(5): 673–93.

Mascarenhas, A. (1977) 'Resettlement and desertification: the Wagogo of Dodoma Distict, Tanzania', *Economic Geography* 53: 376–80.

Mascarenhas, A. and Mbilinyi, M. (1983) *Women in Tanzania: an annotated bibliography*, Uppsala Institute of African Studies, Uppsala.

Maya, R.S. (1988) 'Structural adjustment in Zimbabwe: its impact on women', Unpublished paper commissioned for the Commonwealth Expert Group on Women and Structural Adjustment.

Mbilinyi, M. (1988) 'Agribusiness and women peasants in Tanzania', *Development and Change* 19: 549–83.

Mbilinyi, M. (1989) 'The role of women in promoting local-level development in East Africa', Paper presented to the UNCRD/CIRDAFRICA Seminar on Reviving Local Self-Reliance: Challenges for Rural/Regional Development in Eastern and Southern Africa, Arusha, Tanzania, 24 February.

Mbugua, Moses G. (1989) 'Kenya Freedom from Hunger Council for National Development: experiences in local-level development', Paper presented to the UNCRD/CIRDAFRICA Seminar on Reviving Local Self-Reliance: Challenges for Rural/Regional Development in Eastern and Southern Africa, Arusha, Tanzania, 24 February.

Meyers, R.L. (1981) *Organisation and administration of integrated rural development in semi-arid areas, the Machakos Integrated Development Plan*, Report prepared for the Office of Rural Development and Development Administration, Development Support Division, Agency for International Development, Ithaca, November.

Mikell, G. (1989) *Cocoa and chaos in Ghana*, Paragon House, New York.

Molyneux, M. (1985) 'Mobilization without emancipation? Women's interests, the state, and revolution in Nicaragua', *Feminist Studies* 11(2): 227–54.

Moore, Henrietta (1986) *Space, text and gender: an anthropological study on the Marakwet of Kenya*, Cambridge University Press, Cambridge.

Moore, S.F. (1986) *Social facts and fabrications: 'customary' law on Kilimanjaro 1880–1980*, Cambridge University Press, Cambridge.

Moulaert, F. and Salinas, P.W. (eds) (1983) *Regional analysis and the new international division of labour*, Studies in Applied Regional Science, Boston, The Hague.

Moxon, J. (1984) *Volta – Man's greatest lake*, André Deutsch, London.

Muegge, H. and Stöhr, W. with Hesp, P. and Stuckey, B. (1987) *International economic restructuring and the regional community*, Gower, Aldershot.

Mukoko, R.K. (1987) 'The role of informal sector in rural development: a case of wood carving industry in Wamuyu Location, Machakos District', MA thesis, Department of Urban and Regional Planning, Nairobi University, June.

Mulazi, J.K.N. (1984) *'An evaluation of village vehicle utilization: a case study of selected villages in Dodoma District'*, Unpublished post-graduate diploma thesis, Tanzania.

Munachonga, M.L. (1986) 'Impact of economic adjustment on women in Zambia', Unpublished paper commissioned for UNDP Restructuring and Development in Zambia: Roles for Technical Cooperation Group, New York.

Munslow, B. (1985) 'Prospects for the socialist transformation of agriculture in Zimababwe', *World Development* 13(1): 41–58.

Muro, A. (1987) *Education and training of women for employment: the case of Tanzania*, Arusha, 1982, Eastern and Southern African Management Insitute, Ministry of Education, Dar es Salaam.

Mussa-Nda, N. (1988) 'A greater role for local development strategies', *Regional Development Dialogue* 9(2): 1–11.

Mutizwa-Mangiza, N.D. (1986) 'Local government and planning in Zimbabwe', *Third World Planning Review* 8(2): 153–75.

Mwaluko, E.P. (1961/2) 'Famine in the Central Province of Tanganyika', *Tropical Agriculture* 39, Trinidad.

Naschold, F. (1978) *Alternative Raumpolitik*, Athenian Verlag, Kronburg.

Ndaro, J.M.M. (ed.) (1987) *A survey of small village dam construction projects in Dodoma District*, Institute of Rural Development Planning, Dodoma.

Ndaro, J.M.M. and Temu, J.J. (1989) 'Local strategies and regional/rural development in Kenya: a case study of Machakos District', Paper presented to the UNCRD/CIRDAFRICA Seminar on Reviving Local Self-Reliance: Challenges for Rural/Region Development in Eastern and Southern Africa, Arusha, Tanzania, 24 February.

Ndorobo, B. (1973) 'Famine problems in Dodoma District', *Journal of Geographical Association of Tanzania* 3: 26–47.

Ngugi wa Thiong'o (n.d.) 'Women in cultural work: the fate of Kamiriithu people's theatre in Kenya'.

Nyerere, J.K. (1967) 'The Arusha declaration', pp. 231–50 in J.K. Nyerere (1968) *Freedom and Socialism*, Oxford University Press, Oxford.

Nyerere, J.K. (1968) *Freedom and Socialism: a selection from writings and speeches 1965–7*, Oxford University Press, Oxford.

Nyerere, J.K. (1972) *Decentralisation*, Government Printer, Dar es Salaam.

Oakley, P. (ed.) (1988) *Proceedings of the international symposium on the challenge of rural poverty: how to meet it*, German Foundation for International Development, Feldafing, FGR.

268

Oakley, P. and Marsden, P. (1984) *Approaches to participation in rural development,* ILO, Geneva.

Odingo, R.S. (1988) 'The Eastern African rural settlement pattern and its bearing in local-level development strategy', *Regional Development Dialogue* 9(2): 14–34.

Odingo, R.S. and Okeyo, A.P. (1988) 'The rural development problematique in Eastern Africa', *Regional Development Dialogue* 9(2): i–v.

Odunga, S., Mabele, R.B. *et al.* (1988) 'Tanzania economic trends', *A Quarterly Review of the Economy* 1(1).

OECD (1988) *Activities of the OECD, report by the Secretary General,* OECD Publications, Paris.

O'Keefe, P., Raskin, P. and Bernow, S. (1984) *Energy and development in Kenya: opportunities and constraints,* Beijer Institute, Stockholm, Sweden.

Okeyo, A.P. (1988) 'The role of women in Eastern African rural development', *Regional Development Dialogue* 9(2): 89–101.

Olowu, Dele (1989) 'Local institutes and development: the African experience', *CJAS/RCEA* 23(2): 201–26.

Ondiege, P.O. (1989) 'Local strategies and regional/rural development in Kenya: a case study of Machakos District', Paper presented to the UNCRD/CIRDAFRICA Seminar on Reviving Local Self-Reliance: Challenges for Rural/Regional Development in Eastern and Southern Africa, Arusha, Tanzania, 24 February.

Onimode, B. (1989a) 'Case study on the impact of structural adjustment on women in Nigeria', Unpublished paper commissioned for the Commonwealth Expert Group on Women and Structural Adjustment.

Onimode, B. (ed.) (1989b) *The IMF, the World Bank and the African Debt,* vol. 2, 'The social and political impact', Institute for African Alternatives/Zed Press, London.

Ouma, Stephen O.A. (1989) 'Local strategies and regional/rural development in Eastern and Southern Africa: a report of the Uganda case study', Paper presented to the UNCRD/CIRDAFRICA Seminar on Reviving Local Self-Reliance: Challenges for Rural/Regional Development in Eastern and Southern Africa, Arusha, Tanzania, 24 February.

Parkin, D.J. (1972) *Palms, wine and witnesses: public spirit and private gain in an African farming community,* Chandler, San Francisco.

Patel, D. (1988) 'Some issues of urbanization and development in Zimbabwe', *Journal of Social Development in Africa* 3(2): 17–31.

Pinstrup-Andersen, P., Jaramillo, M. and Stewart, F. (1987) 'The impact of government expenditure', pp. 73–89 in G.A. Cornia, R. Jolly and F. Stewart (eds) *Adjustment with a human face: protecting the vulnerable and promoting growth,* Clarendon Press, Oxford.

Posnansky, M. (1980) 'How Ghana's crisis affects a village', *West Africa* 3306 (December): 2418–20.

Posnansky, M. (1984) 'Hardships of a village', *West Africa* 3506 (October): 2161–3.

Putterman, Louis (1990) 'Village communities, co-operation, and inequality in Tanzania: comments on Collier *et al.*', *World Development* 18(1): 147–53.

Redclift, M. (1984) *Development and the environmental crisis: red or green alternatives?,* Methuen, London.

Redclift, M. (1987) *Sustainable development: exploring the contradictions*, Methuen, London.

Republic of Zimbabwe (1982) *Transitional National Development Plan, 1982/83 – 1984/85*, Harare.

Republic of Zimbabwe (1986) *First Five-Year National Development Plan 1986–1990*, vol. 1, Harare.

Richards, P. (1979) 'Community environmental knowledge in African rural development', *IDS Bulletin* 10(2): 28–36.

Richards, P. (1983) 'Ecological change and the politics of African land use', *African Studies Review* 26(2): 1–72.

Richards, P. (1985) *Indigenous agricultural revolution: ecology and food production in West Africa*, Hutchinson, London.

Robertson, C. and Berger, I. (1986) 'Introduction: analyzing class and gender – African perspectives', pp. 3–24 in C. Robertson and I. Berger (eds) *Women and class in Africa*, Holmes & Meier, New York.

Rohrbach, D.D. (1987) 'A preliminary assessment of factors underlying the growth of communal maize production in Zimbabwe', pp. 145–84 in M. Rukuni and C.K. Eicher (eds) *Food security for southern Africa*, UZ/MSU Food Security Project, University of Zimbabwe, Harare.

Rondinelli, D.A., McCullough, J.S. and Johnson, R.W. (1989) 'Analysing decentralization policies in developing countries: a political economy framework', *Development and Change* 20: 57–87.

Rutashobya, M. (1988) 'Marriage customs and women's legal rights in Tanzania', pp. 29–76 in *Proceedings and papers of the workshop on women and the law: theory and practice*, Dar es Salaam Women Research and Documentation Project, Dar es Salaam.

Sandbrook, R. (1986) 'The state and economic stagnation in tropical Africa', *World Development* 14(3): 319–32.

Sanyal, B. (1984) 'Urban agriculture: a strategy for survival in Zambia', Ph.D. thesis, University of California, Los Angeles; cited in Evans (1989).

Scott, J.C. (1986) *Weapons of the weak: everyday forms of peasant resistance*, Yale University Press, New Haven, Conn.

Sender, J. and Smith, S. (1986) *The development of capitalism in Africa*, Methuen, London.

Senyah, J. (1984) 'Ghana cocoa: more than just a bean', *West Africa* 3483 (May): 1018–19.

Sibanda, B.M.C. (1986) 'Impacts of agricultural microprojects on rural development: lessons from two projects in the Zambezi Valley', *Land Use Policy* 3(4): 311–29.

Sinare, H. (1988) 'Women: do you know your rights? the lack of them?', pp. 17–22 in *Proceedings and papers of the workshop on women and the law: theory and practice*, Dar es Salaam Women Research and Documentation Project, Dar es Salaam.

Smith, S. (1987) 'Zimbabwean women in co-operatives: participation and sexual equality in four producer co-operatives', *Journal of Social Development in Africa* 2(1): 29–46.

Songsore, Jacob (1982) *Co-operative credit unions as instruments of regional development: the example of N.W. Ghana*, Occasional Paper 17, Centre for Development Studies, University College of Swansea, Wales.

Songsore, Jacob (1983) *Intraregional and interregional labour migrations in historical perspective: the case of North-Western Ghana*, Occasional Paper 1, Faculty of Social Sciences, University of Port Harcourt, Nigeria, October.

Songsore, Jacob (1989) 'The ERP/Structural Adjustment Programme and the "distant" rural poor in Northern Ghana', Paper presented at the *International Conference on Planning for Growth and Development in Africa*, ISSER, University of Ghana, Legon, Accra, 13–17 March.

Songsore, Jacob and Aloysius Denkabe (1988) 'Challenging rural poverty in Northern Ghana: the case of the Upper-West region', Unpublished preliminary report for UNICEF, October, UNICEF, Accra.

Stamp, P. (1987) 'Matega: manipulating women's co-operative traditions for material and social gain in Kenya', Paper presented at the Annual Conference of the Canadian Association of African Studies, University of Alberta, Edmonton, May.

Stamp, P. (1989) *Technology, gender and power in Africa*, Technical Study 63e, International Development Research Centre, Ottawa.

Stanning, J. (1987) 'Household grain storage and marketing in surplus and deficit communal farming areas in Zimbabwe: preliminary findings', pp. 245–91 in M. Rukuni and C.K. Eicher (eds) *Food security for southern Africa*, UZ/MSU Food Security Unit, University of Zimbabwe, Harare.

Stöhr, W.B. (1981) 'Development from below: the bottom-up and periphery-inward development paradigm', pp. 39–72 in W.B. Stöhr and D.R.F. Taylor (eds) *Development from above or below?: the dialectics of regional planning in developing countries*, Wiley, Chichester.

Stöhr, W.B. and Taylor, D.R.F. (eds) (1981) *Development from above or below?: the dialectics of regional planning in developing countries*, Wiley, Chichester.

Stöhr, W.B. and Tödtling, F. (1978) 'Spatial equity – some antitheses to current regional development strategy', *Paper of the Regional Science Association* 38: 33–53.

Stoneman, C. and Cliffe, L. (1989) *Zimbabwe: politics, economics and society*, Pinter, London.

Stuckey, B. (1985) *Endogenous development*, NFP 5 Working Papers, Berne.

Sutton, I.B. (1981) 'The Volta River salt trade: the survival of an indigenous industry', *Journal of African History* 22: 43–61.

Swantz, M.L. (1977) *Strain and strength among peasant women in Tanzania*, BRALUP Research Paper 49, University of Dar es Salaam.

Swift, J. (1989) 'Why are rural people vulnerable to famine?'. *IDS Bulletin* 20(2): 8–15.

Taal, H. (1989) 'How farmers cope with risk and stress in rural Gambia', *IDS Bulletin* 20(2): 16–21.

Taasisi ya Watu Wazima Miradi ya Wakina Mama Matetereka (1984) In *Sauti ya Wanawake Tanzania* 1, Idara ya Wanawake, Dar es Salaam.

Taasisi ya Watu Wazima Zijue Haki za Wanawake Tanzania (1987) Dar es Salaam.

Tattersfield, J.R. (1982) 'The role of research in increasing food crop potential in Zimbabwe', *Zimbabwe Agricultural Journal* 16(1): 6–10, 24.

Taylor, D.R.F. (1989) 'Why local initiatives in Africa?: the context and rationale', Paper presented to the UNCRD/CIRDAFRICA Seminar on

Reviving Local Self-Reliance: Challenges for Rural/Regional Development in Eastern and Southern Africa, Arusha, Tanzania, 24 February.

Thiele, G. (1984) 'Location and enterprise choice: a Tanzanian case study', *Journal of Agricultural Economics* 35(2): 257–64.

Thomas, B.P. (1985) *Politics, participation and poverty: development through self-help in Kenya*, Westview, Boulder, Col.

Thomas, B.P. (1988) 'State formation, development, and the politics of self-help in Kenya', *Studies in Comparative International Development* 23(3): 3–27.

Thomas, Helen T. (1988) 'Constraints on rural women's access to credit in the Third World: a review of the issues and literature', pp. 6–19 in F. Mackenzie (ed.) *Gender and processes of change in the Third World*, Discussion papers 7, Carleton University, Ottawa.

Thomas-Slayter, B.P. and Ford, R. (1989) 'Water, soils, food and rural development: examining institutional frameworks in Katheka Sub-location', *Canadian Journal of African Studies* 23(2): 250–71.

Tibaijuka, A. (1988) 'The impact of structural adjustment programmes on women: the case of Tanzania's Economic Recovery Programme', Unpublished paper commissioned for the Commonwealth Expert Group on Women and Structural Adjustment.

Turnbull, M. (1972) *The mountain people*, Jonathan Cape, London.

Turton, D. (1977) 'Responses to drought: the Murai of Southwestern Ethiopia', pp. 165–92 in J.P. Garlick and R.W.J. Keay (eds) *Human ecology in the tropics*, Taylor and Francis Ltd, London.

UN (1986a) *Programme of Action for African Economic Recovery and Development 1986–90* (UN-PAAERD), New York.

UN (1986b) 'The declaration of Non-Governmental Organizations on the African Economic and Social Crisis', Proceedings of a meeting of African, Northern and International NGOs at the UN General Assembly Special Session on the Critical Economic and Social Situation in Africa, 26–31 May 1986, New York.

UNECA (1980) *Lagos Plan of Action and Final Act of Lagos*, Addis Ababa.

UNECA, Organization of African Unity (1986) 'Africa's submission to the Special Session of the UN General Assembly on Africa's economic and social crisis', Addis Ababa, 31 March.

UNECA (1986) *African Priority Programme for Economic Recovery* (APPER), Addis Ababa.

UNECA (1987) *African socio-economic indicators 1985*, UN e/f. 87.11.K.8, Addis Ababa.

UNECA (1989a) *African Alternative Framework to Structural Adjustment Programmes for Socio-economic Recovery and Transformation* (AAF–SAP) Addis Ababa.

UNECA (1989b) *Statistics and policies: ECA*, Preliminary observations on the World Bank report: 'Africa's adjustment and growth in the 1980s', Addis Ababa.

UNECA (1990) *Survey of Economic and Social Conditions in Africa 1987/88*, E/ECA/CM15/3/Rev.2, United Nations, New York.

UNICEF (1984) 'Statistics on children in UNICEF assisted countries', Accra, Unpublished document.

UNICEF (1985) 'Situation analysis of women and children', Accra, Ghana, Unpublished document.

UNICEF (1986) 'United Nations Programme of Action for the Economic Recovery 1986–90, Draft basic document, Preparatory Committee of the Whole for the Special Session of the General Assembly on the Critical Economic Situation in Africa, UNICEF, New York.

UNICEF (1988) *The state of the world's children*, UNICEF Publications, New York.

United Nations Centre for Regional Development (1988) 'Eastern African rural development experience: strategies on local-level development', Report of the Proceedings of an International Seminar, Nairobi, 30 June to 3 July 1988, Nagoya, Japan.

United Nations Centre for Regional Development (1989) *Reviving Local Self-Reliance: Challenges for Rural/Regional Development in Eastern and Southern Africa*, Meeting report series no. 31, Nagoya, Japan.

United Republic of Tanzania (1976) *Third five year plan for economic and social development, 1st July 1976–30th June 1981*, Government Printer, Dar es Salaam.

United Republic of Tanzania (1982) *1978 population census vol. 4: a summary of selected statistics*, Government Printer, Dar es Salaam.

United Republic of Tanzania (1984) *National economic survival programme*, Government Printer, Dar es Salaam.

United Republic of Tanzania (1986) *Economic recovery programme*, Government Printer, Dar es Salaam.

UWT (n.d.) *Taarifa ya jumuiya ya Wanawake ya miaka 20 ya Azimio la Arusha na miaka 10 ya CCM*, Dar es Salaam.

Van Den Dries, J. (1970) *Credit union handbook*, Zenith Printing Works, Nairobi, March.

Van Gelder, B. and Kerkhof, P. (1984) *The agroforestry survey in Kakamega District*, Working Paper 3, Kenya Woodfuel Development Programme, Nairobi.

Vaughan, M. (1985) 'Households units and historical process in southern Malawi', *Review of African Political Economy* 34: 35–45.

Vincent, V. and Thomas, R.G. (1961) *An agricultural survey of southern Rhodesia*, Government Printer, Salisbury.

Wa Diocese (1987) 'Reorganization and motivation of credit unions in Wa Diocese', mimeo, Wa, Ghana, December.

Waal, Alex de (1989) 'Is famine relief irrelevant to rural people?', *IDS Bulletin* 20(2): 63–7.

Wade, R. (1988) *Village republics: economic conditions for collective action in South India*, Cambridge University Press, Cambridge.

Wagao, J. (1988) *Analysis of the economic situation of urban and rural women in Tanzania*, UNICEF, Dar es Salaam.

Watts, Michael (1986) 'Geographers among the peasants: power, politics and practice', *Economic Geography* 2: 373–86.

Watts, Michael (1988) 'Idioms of land and labour; producing politics and rice in Senegambia', Paper presented at the Symposium on .Land in African Agrarian Systems held at the University of Illinois, Urbana-Champaign, 10–12 April.

Watts, Michael (1989) 'The agrarian crisis in Africa: debating the crisis', *Progress in Human Geography* 13(1): 1–41.

Weaver, C. (1984) *Regional development and the local community: planning, politics and social context*, Wiley, Chichester.

Weiner, D. (1988) 'Agricultural transformation in Zimbabwe: lessons for South Africa after apartheid', *Geoforum* 19(4): 479–96.

Wekwete, K.H. (1988) 'Rural growth points in Zimbabwe: prospects for the future', *Journal of Social Development in Africa* 3(2): 5–16.

Were, S. and Akong'a, J. (eds) (1981) 'Machakos District social–cultural profile', draft report for the Ministry of National Planning and Development, Institute of African Studies, University of Nairobi, Nairobi.

West Africa (1989) November, West Africa Publishing Company Ltd, London.

White, B. (1976) 'Population, involution and employment in rural Java', *Development and Change* 7(3): 267–90.

Whitlow, J.R. (1980a) 'Land use, population pressure and rock outcrops in the tribal areas of Zimbabwe Rhodesia', *Zimbabwe Rhodesia Agricultural Journal* 77(1): 3–11.

Whitlow, J.R. (1980b) 'Environmental constraints and population pressures in the tribal areas of Zimbabwe', *Zimbabwe Agricultural Journal* 77(4): 173–81.

Whitlow, J.R. (1988) *Land degradation in Zimbabwe: a geographical study*, Natural Resources Board, Harare.

Wilks, I. (1977) 'Land, labour, capital and the forest kingdom of Asante: a model for change', pp. 487–534 in J. Friedman and M. Rowlands (eds) *The evolution of social systems*, Duckworth, London.

World Bank (1981) *Accelerated development in Sub-Saharan Africa: an agenda for action*, Washington, D.C.

World Bank (1984a) *Toward sustained development in Sub-Saharan Africa: a joint program of action*, Washington, D.C.

World Bank (1984b) *World development report*, Washington, D.C.

World Bank (1985) *World development report*, Washington, D.C.

World Bank (1986) *World development report*, Washington, D.C.

World Bank (1988a) *Ghana: country economic memorandum*, International Bank for Reconstruction and Development, Washington, D.C., November.

World Bank (1988b) *Report on adjustment lending*, Washington, D.C.

World Bank (1989) *Sub-Saharan Africa: from crisis to sustainable growth*, Washington, D.C.

World Bank/UNDP (1989) *Africa's adjustment and growth in the 1980s*, Washington, D.C.

World Bank (1990) *World development report 1990*, Oxford University Press, Toronto.

World Commission on Environment and Development (1987) *Our common future* (the Brundtland Report), Oxford University Press, Oxford.

Young, R. (1988) *Zambia: adjusting to poverty*, North–South Institute, Ottawa.

Young, Roger (1989) 'Structural adjustment in Africa: roles and challenges for NGOs', Paper presented to Canadian Council for International Co-operation, May, Ottawa.

Zinyama, L.M. (1986a) 'Agricultural development policies in the African farming areas of Zimbabwe', *Geography* 71 (part 2): 105–15.

Zinyama, L.M. (1986b) 'Rural household structure, absenteeism and agricultural labour: a case study of two subsistence farming areas in Zimbabwe', *Singapore Journal of Tropical Geography* 7(2): 163–73.

Zinyama, L.M. (1987a) 'Assessing spatial variations in social conditions in the African rural areas of Zimbabwe', *Tijdschrift voor Economische en Sociale Geografia* 78(1): 30–43.

Zinyama, L.M. (1987b) 'Gender, age and the ownership of agricultural resources in the Mhondoro and Save North communal areas of Zimbabwe', *Geographical Journal of Zimbabwe* 18: 1–14.

Zinyama, L.M. (1988a) 'Commercialization of small-scale agriculture in Zimbabwe: some emerging patterns of spatial differentiation', *Singapore Journal of Tropical Geography* 9(2): 151–62.

Zinyama, L.M. (1988b) 'Farmers' perceptions of the constraints against increased crop production in the subsistence communal farming sector of Zimbabwe', *Agricultural Administration and Extension* 29(2): 97–109.

Zinyama, L.M. (1988c) 'A comparative analysis of social and economic factors influencing agricultural change and development in the Mhondoro and Save North communal areas of Zimbabwe', Unpublished D.Phil. thesis, Department of Geography, University of Zimbabwe, Harare.

Zinyama, L.M. (1989) 'Local strategies and responses and their contribution to rural development in Zimbabwe', Paper presented to the UNCRD/ CIRD AFRICA Seminar on Reviving Local Self-Reliance: Challenges for Rural/Regional Development in Eastern and Southern Africa, Arusha, Tanzania, 24 February.

Zinyama, L.M. and Whitlow, R. (1986) 'Changing patterns of population distribution in Zimbabwe', *GeoJournal* 13(4): 365–84.

Zinyama, L.M., Campbell, D.J. and Matiza, T. (1990) 'Land policy and access to land in Zimbabwe: the Dewure resettlement scheme', *Geoforum* 21(3): 359–70.

INDEX